MARINE CHEMISTRY

MARINE CHEMISTRY

(IN TWO VOLUMES)

DEAN F. MARTIN

DEPARTMENT OF CHEMISTRY AND
INSTITUTE OF MARINE SCIENCE
UNIVERSITY OF SOUTH FLORIDA
TAMPA, FLORIDA

VOLUME 1

Analytical Methods

SECOND EDITION
REVISED AND EXPANDED

MARCEL DEKKER, INC., New York

1972

MARCEL DEKKER, INC.
95 Madison Avenue, New York, New York 10016

LIBRARY OF CONGRESS CATALOG CARD NUMBER 68-27532

ISBN: 0-8247-1456-3

PRINTED IN THE UNITED STATES OF AMERICA

PREFACE

As I noted in the Preface to the First Edition, the field of marine chemistry can be divided into two subject areas: theory and analytical methods. Marine chemistry is, and for the foreseeable future will continue to be, an experimental science, with considerable effort being given to the collection of water samples and the analysis of constituents. The information that results from this effort forms the basis or the support of the theories of marine chemistry and also functions as a major service of the field to those in allied fields. For this reason, it seemed logical to initiate the present work by first considering the analytical methods, and then in a second volume covering the results and theoretical aspects that follow from experimental observations.

In this second edition, I have attempted to follow the plan of the first edition: to prepare a book for the typical new student or investigator in marine chemistry who does not have professional experience in the field. Typically, these persons have had a year of general chemistry, but it may often safely be assumed that they would benefit from a review of the basic concepts of general chemistry that now are obviously pertinent. Typically, also, it may be assumed that these persons are experienced in the laboratory, though not necessarily in the chemistry laboratory.

I have also tried to follow the initial organizational plan of the first edition, though numerous corrections of detail and additional methods have been introduced, and the organizational plan may be more obviously divided into five parts. For each determination, as before, an introduction is presented, and this is followed by one or more experimental methods, each of which is divided into several subsections (reagents, equipment, sampling and storage, experimental methods, calculations); notes and references follow the experimental sections. As in the first edition, choices are given where possible and appropriate. For example, in the determination of chlorinity the classical Knudsen procedure is given, but it is paralleled by a recent potentiometric procedure. Also, the determination of nitrate by a convenient, routine cadmium reduction procedure is given, but it is followed by the elegant Armstrong procedure, which is eminently useful in studies of culture systems.

The features that remain from the previous edition and those which have been added to this one include the following:

1. Theory and practice are presented as an integrated whole. A substantial review of basic concepts and methods is presented in the first section. Pertinent calculations and concepts are reviewed in Chapter 2 and are keyed to the determinations that follow. A review of acids, bases, and pH (Chapter 3) precedes the determination of alkalinity, total carbonate, and total carbon dioxide (Chapter 4). The second section (Analysis of selected micronutrients and related compounds) is initiated with a discussion of spectrochemical analysis (Chapter 8) which serves as a useful empirical basis for the twelve chapters that follow. In the fourth section (Analysis of polluted waters), the opening chapter (Chapter 24) keys the determinations and the section closes with a review of the mercury problem and methodology (Chapter 32).

2. The volume is written for potential users who have widely differing backgrounds. Each chapter begins with a review of pertinent methods, and the reader is presented with a series of alternatives. I do not believe that it is essential or even desirable to acquaint the user with each determination or variation that he is likely to encounter. I trust that a suitable and thorough background for some less common determinations has been provided, and I believe that this is pedagogically sounder and more realistic. For each determination, a set of clear, specific directions are provided that depart from a "cookbook" approach to a fair degree.

3. This volume includes most of the commonly used determinations of marine chemistry, as well as some less commonly used ones that are likely to be significant (direct determination for boron, urea, arsenic, polyphosphate, and mercury, among others). Many of the problems of analysis are considered on a start-to-finish basis. Problems of sampling, storage, and order of analysis are considered in Chapter 1. Details of data treatment are reviewed in Chapter 2, and specific problems are subsequently considered in individual chapters. It is impossible to cover every

problem in the sampling-to-reporting chain, and I make no claim to have done so. I should like to think, however, that the major problems and numerous minor ones have been covered and that the unforeseen ones will be communicated to me.

4. This volume includes, as the previous edition did, some of the newer methods of analysis that seem to be omitted from available manuals of marine chemistry. One of these methods is flame photometry (Chapter 23), which is not new, but neither is it commonly covered. Atomic absorption spectroscopy has been used with considerable success in trace-metal analysis and an introduction to the method and its application to marine chemistry is given in Chapter 22. Also, a modification of this latter method, flameless atomic absorption spectroscopy, is covered in Chapter 32 in the discussion of the determination of mercury. Compleximetric titrations have been used many years for calcium and magnesium analyses, but the modifications used here (Chapter 21) represent improvements in methodology. Specific-ion electrodes seemingly have not been widely applied to analysis of constituents of sea water, but the analysis of fluoride by means of a fluoride electrode is a notable exception (Chapter 30).

5. Several significant determinations have been added to this edition, as a result of improved methodology. These include determinations for urea, molybdenum, arsenic, hydrogen sulfide, polyphosphate, and mercury. A section on analysis of sediments has been expanded to include related determinations (humic acids, total organic carbon content, and nucleic acids).

I recognize the impossibility of providing a set of determinations together with background and theory that will provide all users with everything they consider desirable. Many workers are especially fortunate in the facilities and equipment available to them, but the majority are faced with more limited resources, and this volume is aimed at this second group. I hope also that many who use this volume will have helpful suggestions; a number of the suggestions made by users of first edition are incorporated into this edition. I shall welcome all such suggestions.

I am deeply grateful to many students, colleagues, and friends for helpful criticism, and to members of the administration at the University of South Florida for advice and support. I am particularly grateful to my wife, Barbara B. Martin, who provided me with substantial technical assistance. Finally, I am especially grateful to Mrs. Aura Ferrell who typed the manuscript with the especial care that is necessary for the process used in this edition.

DEAN F. MARTIN

TO W. H. T.

CONTENTS

MARINE CHEMISTRY

Section I

BASIC CONCEPTS AND METHODS

1

INTRODUCTION TO MARINE CHEMISTRY

1.1 Introduction

The marine chemist is faced with a variety of problems which seem to fall into four groups: nature of the sample, the number of samples, the number and kind of analyses required, and the conditions under which the analyses are performed. The sample may be sea water, or estuarine water of lower salinity, or ground water. The sample may be a sediment or an organism, the composition of which is of interest. The water sample may be contaminated with pollutants, in which case alteration of the analysis procedure is needed. These and other problems in the first group are considered in the chapters that follow.

The second group of problems, associated with the number of samples, has arisen from the studies of those interested in physical and biological aspects of natural waters. The large number and high frequency of sampling for many of these studies has provided an impetus toward several changes; these include *in situ* analysis (pH, salinity, oxygen content) which obviate the need for sample removal, rapid methods of analysis (for example, atomic absorption spectroscopy) and automation (for example, the Technicon Auto-Analyzer which Henriksen[1] has used to analyze the nitrate content of 27 samples per hour). The first and third of these developments has received (slight) treatment here; the former, because directions accompanying the particular instrument are usually models of clarity; the latter, because the present cost of such instrumentation may be a limiting factor.

The third group of problems--the number and kind of analyses--are closely related to the nature of the sample. As Culkin[2] notes, "the composition of sea water poses some difficult problems." At first glance at Table 1-1, the sea water

TABLE 1-1

Composition of Sea Water

Major Constituent	Concentration[a] (parts per million)
Chloride	19,353
Sodium	10,760
Sulfate	2,712
Magnesium	1,294
Calcium	413
Potassium	387
Bicarbonate	142
Bromide	67
Strontium	8
Boron	4
Fluoride	1

[a]Mg/kg of sea water of salinity $35^{o}/oo$[2].

appears to be an ideal system of an aqueous solution of 11 major (or "conservative") constituents (Note 1) with relatively constant composition. This is misleading on several counts. First, as Carpenter and Carritt[3] noted, the tendency is to regard the relative composition as constant in spite of the local variations that have been documented. Second, many analyses of a given constituent are subject to considerable hazards because of interference by other ions; for example, there is as yet no satisfactory procedure for analyzing the major constituent, sodium. Third, the concentrations of many elements of interest, the so-called minor constituents, are so low as to impose special problems.

Finally, the fourth group of problems--the analysis conditions--must be considered. There are considerable difficulties in carrying out analyses at sea, but many problems of the past have been overcome through ingenious solutions. The effects of ship motion on the analyst and the instrument can be compensated to some extent, but difficulties remain. Precision weighing is impossible and some instrumental techniques (e.g., polarography) cannot be used on typical vessels. On the other hand, a wide range of techniques is available, including titrimetry, potentiometry, pH measurements, and photometry (most recently, atomic absorption spectroscopy). Many problems of contamination are magnified at sea, but these can be overcome with care, once they are recognized.

Certain problems remain. Most vessels do not have facilities that match land-based laboratories, and some believe that if highest precision is required, the samples must be brought to shore laboratories for analysis. This requirement adds the difficulties of storage of water samples to those of sampling and treatment of samples. These problems are considered in subsequent sections.

1.2 Water Sampling

Until new materials or equipment become available, the marine chemist must make do with water sampling devices that have not changed in principle since the Hooke invention in the seventeenth century. The satisfactory water sampling device must collect a large, representative water sample at a known depth and bring it unaltered and uncontaminated to the surface. It is safe to say that, in the strict sense, a truly satisfactory water sampling device does not exist because of the uncertainties of contamination and depth and difficulties inherent in the method. The types of samplers available have been reviewed by Herdman[4] and Riley.[5] Practical details of water sampling will not be described because they are obtainable elsewhere.[6]

The Hydrocast. Deep-water samples are usually collected with two kinds of water sampling devices: The reversing sampler, such as the Nansen bottle, or the nonreversing sampler (with a reversing thermometer rack) such as the Niskin bottle.

The bottles are attached to a weighted hydrographic wire (hydrowire) at appropriate intervals (Table 1-2) and lowered from a platform at the side of the vessel. The samplers are open at the bottom and the top so that they are flushed with sea water as they are lowered. The Nansen bottles have poor flushing characteristics because the openings are narrow; the Niskin bottles have excellent flushing characteristics. When the bottles are equilibrated at the preselected depths, it is necessary to trip the device that closes the top and bottom openings. This is done with a messenger, a brass or lead weight, which is attached to the hydrowire and dropped from the hydrocast platform. The messenger trips the closing device and releases a second messenger which slides down the hydrowire

TABLE 1-2

The Standard Depths in Meters

0	300	2000
10	400	2500
20	500	3000
30	600	4000
50	(700)	5000
75		
100	800	6000
150	1000	7000
200	1200	etc.
(250)	1500	

to trip the closing device and release a third messenger, and so on. Thus an uncontaminated sample is obtained, though the valves on the Nansen sampler do not seal perfectly on closing.

The temperature and depth are determined by means of protected and unprotected reversing thermometers. These are delicate and expensive thermometers which can be read to 0.01°. When the messenger trips the sampler, the thermometers are reversed by 180°. This disconnects the mercury column from a mercury reservoir, and by a series of corrections, it is possible to determine the water temperature *in situ* at the time of reversal. The protected thermometer is protected from the effect of hydrostatic pressure and gives the true temperature. In contrast, the unprotected thermometer is in direct contact with sea water and is affected by hydrostatic pressure. The readings of the two thermometers differ by about 1° for each 100 meters of depth. Despite the delicacy of the thermometers, the percentage of malfunction is surprisingly low. Thorough discussions of the use of reversing thermometers may be found elsewhere.[6,7]

Selection of Depths. In 1936, the International Association of Physical Oceanography proposed the set of standard depths listed in Table 1-2. Beyond the 4000-meter level, the standard depths occur at every 1000 meters to the bottom. Optional values are given in parentheses. It was proposed that observations should be taken at the standard depths or (more realistically) the data should be adjusted by interpolation from the values at nonstandard depths (Table 1-2). This is reasonable enough, but there are several difficulties.

1.3 Some Problems of Water Sampling

A number of problems are associated with water sampling, including the following.

Lack of Information. Lewis and Goldberg[8] noted in 1954 that analyses in the deep sea were rare. Only 4 out of 39 investigators had obtained samples at depths greater than 1000 meters, though 90% of the oceanic waters are below 1000 meters.

Weight Limitation of the Hydrowire. The number of sampling bottles that can be attached is limited. For example,[4] a 4-mm-diameter wire cord (breaking load 20 cwt) with a 100-ℓb weight at the end would have a safe working load at 5000 meters depth at 750ℓb (or about one-third the breaking load). Because the weight of the hydrowire in water will be 630 ℓb, the weight of the sampling bottles must be less than 120 ℓb. This means that when the weather conditions are very good, the maximum number of water samplers is 12 (Knudsen, Fjarlie) to 6 (Ekman). Obviously, two or three casts may be needed to get adequate samples.

Need for Larger Samples. This is an ever-increasing problem as more analyses or analyses of certain trace elements are required. The Van Dorn samplers[4,5] have been used to collect up to 50 liters. Samplers for collecting volumes up to

220 liters for analysis of fission products have been described by several workers.[5,9] Finally, Schink[10] developed procedures for collecting massive volumes (up to 50 tons).

Problems of Collecting Sterile Water Samples. These problems and the appropriate devices have been described by Niskin.[11]

Time Requirements. There is an inherent difficulty in the hydrocast method that the vessel must be stationary for the time needed to complete the hydrocast. Drifting of the vessel and resultant large wire angles can be limited by use of a bow thruster. Nevertheless, time must be allowed for the bottles to reach equilibrium (a minimum of six minutes), for the messenger to reach and trip the sampling device (three minutes at 600 meters), and for the bottles to be brought up and removed. Several automatic sampling devices have been described[4,5] which permit limited sampling from a moving vessel.

Contamination of the Sample. Few persons can appreciate the many ways in which contamination can occur. This is a critical problem with accurate analyses of trace constituents. Probably the major discrepancies in reported values of trace constituents are due to different modes of sample contamination. Some examples which have been noted[5,12,13] follow.

Obviously, metallic samplers should not be used if the water is to be analyzed for such metals as copper, lead, and zinc. Less obviously, the oxygen content may change if the sample remains in contact with a brass sampler for periods of an hour, which is not unusual for deep hydrocasts. Sampling with a plastic water sampler seems essential. Such devices have been described by Herdman[4] and Niskin (Note 2); these are equipped with thermometer frames for three thermometers. These devices seem to offer a better solution to the problem of metallic contamination than the use of nylon liners or silver or nickel plating, which crack or deteriorate.

The samples may also be contaminated by the corroded hydrowire, with iron and zinc the most likely contaminants. Analysis of organic constituents is particularly subject to contamination by oil from hydrowires.

Next, the sample can become contaminated in transfer from the sampler to the container. Contamination from the hands of operators (rust, copper, amino acids, and other organic compounds) or by ash (smoke, lighted cigarettes) are common hazards.

Finally, the sample may become contaminated by the container, a problem that will be considered in the next section.

Thermometer Problems. Occasionally, as noted before, deep-sea reversing thermometers do seem to malfunction and there is poor agreement between depth readings on adjacent bottles on a given hydrocast and uncertainties of as much as 40 meters. Some malfunctions can be attributed to the fact that these thermometers are "a tribute to the glassblower's art" rather than precision instruments.[4] On the other hand, under optimum conditions, a mean error of 2 meters at 2000 meters is possible.[14] Cooper[14] has shown that many anomalies can be explained in terms of vertical yawing while a messenger is running between bottles. This problem arises during non-optimum conditions and is of serious concern because temperature and salinity do not correspond. In the past, errors were not as apparent because of less precise salinity determinations. Finally, a gauge, developed by Carruthers,[15] can give information on the direction and slope of the hydrowire; with thermometric methods, only the latter can be inferred.

1.4 Storage of Samples

After a sample of water has been brought to the deck in the water sampler, the problem of storage inevitably arises. One exception is the determination of dissolved oxygen content; the samples for this determination are removed and the analysis started before other samples are withdrawn (Chapter 7). The consideration of storage involves three areas: separation of particulate material, selection and treatment of the storage container, and storage. These are considered in the sections that follow.

Separation of Particulate Matter. Natural waters contain varying amounts of suspended particulate matter. This material must be separated as soon as possible after collection, for two reasons. First, the composition of the sample may change as trace constituents can be lost from the sample by adsorption on particulate material. Second, the composition of the particulate material may change as trace constituents are lost.

Filtration by membrane-filtering apparatus (Note 3) has proven to be most satisfactory and is the generally accepted procedure. There is an operational advantage--the material that is retained by a 0.5-μ pore size filter is accepted as being particulate matter for various determinations (particulate iron and copper, and organic phosphorus). Moreover, the membrane filters have other advantages. The pore size is highly uniform and reproducible, filtration is rapid, and the degree of contamination is minimal.

Samples for three determinations are not subjected to filtration analysis; dissolved oxygen, pH, and alkalinity.

The Storage Container. The successful container obviously must prevent loss of water through evaporation, contamination through leaching, or loss of constituents by adsorption. At present, the choice of container is twofold; hard glass or hard (high-density) polypropylene. Both have their limitations. Before either type is selected for storage of samples for trace-constituent analysis, it must be tested to determine whether the test constituent is adsorbed or increased. This can be done with a standard sample, but many prefer using radio-tracer methods.[5,14] The following is an indication of some of the problems.

Generally, hard glass is suitable for storage of samples for salinity and for analyses of many trace constituents. Bakelite stoppers with polyethylene inserts are satisfactory closures, but rubber stoppers or glass stoppers should not be used, because of contamination. Cooper[14] suggested the use of hard glass vials similar to those in which standard sea water is supplied, which could be sealed at once on board with a portable blow-pipe flame (Note 4). The problems of leakage and contamination would be nil if this suggestion were followed. Even so, there may be problems with certain metals. Manganese seems to be leached slowly from borosilicate glass and lead (nitrate) solutions tend to plate out on glass unless the solutions are quite acidic.[12] Appreciable amounts of lead, zinc, and arsenic can be leached from borosilicate glass.[16] Possibly the answer to these problems is pretreatment. The glassware can be soaked with dilute hydrochloric acid prior to use.

Even with these precautions, glass containers are unsuitable for samples to be analyzed for sodium (and probably other alkali metals) and silicate. For this reason, hard polypropylene has been used for the storage of these samples.

The advantages of polypropylene have been recognized; the difficulties are less prominent. Soft polyethylene bottles tend to "breathe" and evaporation can occur.

New polyethylene containers must be treated with dilute hydrochloric acid to remove traces of metal left by the manufacturing process. For example, Matson and Roe[17] reported contamination of solutions with copper leached from the polyethylene container; the resulting concentrations far exceeded the concentration of copper in sea water.

Washing plastic containers with acid must be done judiciously. Nitric acid has converted some plastic surfaces to efficient ion-exchangers; up to 10^{-12} mole of trace metals can be removed per square centimeter of surface.[17]

Phosphate ion "disappears" from filtered solutions stored in polyethylene containers.[18] There is some disagreement about the reason for the disappearance.[5] Heron[18],who attributes the disappearance to uptake by bacteria on the surface of the container, has devised a treatment that makes polyethylene bottles suitable for phosphate samples.

Some plastic containers introduce organic substances into the sample; these may interfere in spectrophotometric determinations.[12]

Special containers (e.g., glass vessel with mercury seals) are used for samples to be analyzed for dissolved nitrogen or inert gases.

Storage. Having selected a suitable storage container, the next problem is selecting the suitable conditions. These are best considered in order of priority.

Samples for dissolved oxygen (Chapter 7) are removed first, treated with Winkler reagents, and stored in alkaline conditions. Glass-stoppered BOD bottles (Chapter 24) are especially useful because the stoppers can be covered with water to prevent air from entering the bottle.

Next, analyses of micronitrients (Chapter 8) should be started within an hour, before changes occur (decomposition or adsorption). If this is not possible, the sample may be stabilized for a few hours (by freezing to 0°C) or for a few months (by sudden cooling to -20° and storage at this temperature). This treatment is effective for nitrate, nitrite, and, probably, silicate. Filtered phosphate samples can be stabilized for several weeks by the addition of chloroform, provided the samples are stored in the dark.

Trace metal elements can disappear with appalling ease during storage. Adsorption of these elements can be minimized by acidifying the sample (typically to pH 2). This can be done conveniently by placing a known volume of purified concentrated hydrochloric acid (Note 5) in a hard polypropylene container. Obviously, it is necessary to test that this treatment does not introduce an impurity in the form of a test element.

Finally, it is useful to remember the Turkish proverb, "One should not descend into a deep well on an old rope." This, of course, implies the need to check the purity of all reagents. Perhaps, more to the point, it implies the need to review the pertinent calculations in marine chemistry. This is done in the next chapter before considering the analytical procedures.

NOTES

1. There is reason to doubt[2] that fluoride is a conservative constituent, and that there is a substantial variation in the fluoride/chlorinity ratio with depth. Chlorinity is discussed in Chapter 5.
2. These are almost completely constructed of plastic. Niskin samplers are available from CM^2, Mountain View, Calif. Van Dorn water samplers are available in a variety of sizes from Hydro Products, San Diego, Calif.
3. One such apparatus is available from the Millipore Filter Corporation, Inc., Bedford, Mass.
4. An example is the portable propane-oxygen Microflame Torch, Microflame, Inc., Minneapolis, Minn.
5. The acid should be passed through a cation exchange resin, and then an anion exchange resin before using (cf. Chapter 12).

REFERENCES

1. A. Henriksen, Analyst, 90, 83 (1965).
2. F. Culkin, in Chemical Oceanography (J. P. Riley and G. Skirrow, eds.), Vol. 1, Academic Press, New York, 1965, p. 122.
3. J. H. Carpenter and D. E. Carritt, Nat. Acad. Sci., Nat. Res. Council Publ., 600, 67 (1959).

4. H. F. P. Herdman, in The Sea (M. N. Hill, ed.) Vol. 2, Interscience, New York, 1963, pp. 124-127.

5. J. P. Riley, in Chemical Oceanography (J. P. Riley and G. Skirrow, eds.), Vol. 2, Academic Press, New York, 1965, pp. 295-304.

6. Instruction Manual for Oceanographic Observations, H. O. Pub. No. 607, 3rd ed., U. S. Hydrographic Office, Washington, 1968.

7. H. U. M. Sverdrup, M. W. Johnson, and R. H. Fleming. The Oceans: Their Physics, Chemistry, and General Biology, Prentice-Hall, New York, 1942.

8. G. L. Lewis, Jr. and E. D. Goldberg, J. Mar. Res., 13, 183 (1954).

9. R. H. Bedman, L. V. Slabaugh, and V. T. Bowen, J. Mar. Res., 19, 141 (1961).

10. D. R. Schink, Thesis, Scripps Institution of Oceanography, University of California, LaJolla, 1962.

11. S. L. Niskin, Deep-Sea Res., 9, 501 (1962).

12. D. N. Hume, Adv. Chem. Ser., 67, 30 (1967).

13. L. H. N. Cooper, J. Mar. Res., 17, 130 (1958).

14. L. H. N. Cooper, in Oceanography (M. Sears, ed.) Publ. 67, A.A.A.S., Washington, 1961, pp. 599-604.

15. J. N. Carruthers, J. Mar. Res., 17, 113 (1958).

16. E. B. Sandell, Colorimetric Determination of Traces of Metals, 3rd ed., Interscience, New York, 1959.

17. W. R. Matson and D. K. Roe, Analysis Instrumentation, 1966, Plenum Press, New York, 1966.

18. J. Heron, Limnol. Oceanogr., 7, 316 (1962).

2

CALCULATIONS IN MARINE CHEMISTRY

2.1 Introduction

The importance of the quantitative aspects of marine chemistry cannot be overemphasized. Because of this, considerable emphasis must be placed on the solving of problems. Many of these problems are types that have been encountered before or represent extensions of previous problem types. For this reason, it seems appropriate to review these problem types through the use of model examples in the sections that follow. Commonly, the problems of marine chemistry involve an understanding of significant figures, exponential numbers, logarithms, and precision. These topics will be reviewed briefly before considering the model examples.

2.2 Significant Figures (Note 1)

Every measured quantity in marine chemistry possesses an uncertainty which is controlled by the limits of precision of measurement. The precision of a result which involves several measurements can be no better than the precision of the measuring instrument which is the least precise. Because of this limitation, the numerical result should reflect no greater precision than is justified by the individual measurements.

For example, recording the weight of a sample as being 10.3754 g would imply that the sample was weighed on an analytical balance. It would imply that the weight was known to the nearest milligram with certainty and that the figure 4 was uncertain. The significant figures are those that are known with certainty and the last figure is always the estimated one. In this example, there are six significant figures, and the recorded result implies that the result lies between 10.3753 and 10.3755 g (or a degree of error of one part in 103,754).

There is a special problem with the figure zero because it is used in two ways: to represent an amount and to fix the decimal place. When it is used in the former way, it is a significant figure; when it is used in the latter way, it is not. Zeros are always significant figures when they appear between other digits. For example, the number 10.0004 g has six significant figures. In all other instances, zeros are significant figures only if they represent measured quantities.

For example, the weight 1200 g has two significant figures, with the zeros being used to fix the decimal point. The result would be better expressed as 1.2 kg. On the other hand, the weight 120.0 g contains four significant figures because it implies that the actual result lies between 119.9 and 120.1 g.

2.3 Exponential Numbers

The extremely small and extremely large numbers that are

commonly encountered in marine chemistry are cumbersome in calculations and difficult to interpret in terms of significant figures. For example, the concentration of fluoride ion is 0.00007 moles/liter of sea water; and the total volume of the ocean has been estimated as 1,370,000,000,000,000,000,000 liters. These numbers can be expressed as a logarithm, or better as an exponential number as a product involving 10 to an appropriate power. Thus the above examples would be written as 7×10^{-5} moles/liter and 1.37×10^{21} liters.

In the first example, the decimal point was moved five places to the <u>right</u>, which meant that the quantity had to be multiplied by 10^{-5}. In the second example, the decimal point was moved 21 places to the <u>left</u>, which meant that the quantity had to be multiplied by 10^{21}.

Calculations are immensely easier when exponential numbers are used.

Multiplication involves the algebraic addition of exponents. For example, the total amount of fluoride in the ocean would be

$$7 \times 10^{-5} \frac{\text{moles}}{\text{liter}} \times 1.37 \times 10^{21} \text{ liters} = 7 \times 1.37 \times 10^{(-5 + 21)}$$
$$= 9.59 \times 10^{16} \text{ moles}$$

Conversely, division requires the algebraic subtraction of exponents.

2.4 Logarithms

The common or Briggsian logarithm of a number is defined as the power to which 10 must be raised to equal that number (Note 2). This is expressed as $\log_{10}$ or log. For example, the logarithm of 1000 (or 10^3) is 3; the logarithm of 1 (or 10^0) is 0; and the logarithm of 0.001 (or 10^{-3}) is -3.

Most of the numbers we encounter are not exact multiples of 10, so it is important to note some additional points about logarithms.

There are two parts of the logarithm: the characteristic and the mantissa. Consider a number written in the exponential form: $A \times 10^n$, where A has a value between zero and ten. The characteristic is n, a zero or an integer, placed to the left of the decimal point in the logarithm. For numbers greater than zero, the characteristic is positive; for numbers less than zero, the characteristic is negative. The mantissa is the numerical value listed in the table of logarithms for A. The mantissa is written to the right of the decimal point in the logarithm and is always positive. The following examples should illustrate these points:

$$\log 47015 = \log(4.7015 \times 10^4) = 4.67224$$

$$\log 4701.5 = \log(4.7015 \times 10^3) = 3.67224$$

$$\log 470.15 = \log(4.7015 \times 10^2) = 2.67224$$

$$\log 47.015 = \log(4.7015 \times 10^1) = 1.67224$$

$$\log 4.7015 = \log(4.7015 \times 10^0) = 0.67224$$

$$\log \quad 0.47015 = \log(4.7015 \times 10^{-1}) = \bar{1}.67224$$
$$= 0.67224 - 1$$
$$\log \quad 0.0047015 = \log(4.7015 \times 10^{-3}) = \bar{3}.67224$$
$$= 0.67224 - 3$$

Logarithms, like exponents, are added in multiplication and subtracted in division. The square root (or the nth root) of a number obtained by taking one-half (or 1/n) times the common logarithm and obtaining the final answer from a table of logarithms. Raising a number to the nth power is effected by multiplying the common logarithm by the value of n and obtaining the final answer from a table of logarithms.

Some confusion arises when the characteristic is negative. Referring to the previous example, it is apparent that $\bar{3}.67224$ is awkward in computations because only the characteristic is negative. An alternative form, 0.67224 -3, is more useful in calculations, but in order to use the logarithm table, the first or second forms must be used.

The third form is also useful in the p(x) notation. For any quantity x, we define p(x)

$$p(x) = -\log_{10}(x) \tag{2-1}$$

and obtain a number whose positive magnitude increases as x becomes smaller. To illustrate,

$$pH = -\log(H^+) \tag{2-2}$$

$$pAg = -\log(Ag^+) \tag{2-3}$$

$$pK_B = -\log K_B \tag{2-4}$$

For aqueous ammonia, the value of K_B is 1.8×10^{-5} mole/liter, and the value of pK_B is 4.7.

2.5 Precision

Unfortunately, the terms precision and accuracy are often used interchangeably. "Precision" is the narrowness of limits between which it is believed that the true value of a measurement or a determination lies. It is indicated by the agreement between successive determinations and is a measure of the reproducibility of those determinations. "Accuracy" is the agreement between a determined value and the true value. Because the true value often cannot be ascertained with absolute certainty, the agreement implied by the accuracy must be estimated, often from a statistical analysis of the measurements and the precision. Much of this depends upon a thorough study of errors that are encountered in quantitative analysis.

These include "determinate errors" (instrumental errors, apparatus errors, personal errors, such as color insensitivity, and errors inherent in the method), which are usually consistent from one determination, and "indeterminate errors", which seem to occur for no apparent cause, even in the hands of an able analyst. Corrections can be applied to compensate for the first type of error (through running a blank or running a control). Corrections cannot be applied to compensate for the second type of error, but an intelligent analysis of the results can usually lead to the selection of the "best" value.

In general, it may be said that large errors are unlikely; small errors occur much more frequently than large ones; and for a given size of error, the number of positive errors balances the number of negative errors.

There are various ways of expressing errors and indicating an estimate of the validity of a set of measurements. Two of these ways are the average deviation and the standard deviation.

Average Deviation ($\bar{a}$). This is the simple arithmetic average of the individual deviations (from the mean), taken without regard to sign:

$$\bar{a} = (|d_1| + |d_2| + d_3 + \ldots + d_n)/n \qquad (2\text{-}5)$$

where $\bar{a}$ is the average deviation and n is the number of measurements in the set, and d_1 is the absolute deviation,

$$|d_1| = |X_1 - \bar{X}| \qquad (2\text{-}6)$$

X_1 and $\bar{X}$ are an individual measurement and the arithmetic mean, respectively. The average deviation is a useful expression because it gives an estimate of the validity of an individual measurement in terms of the deviation that is likely associated with it. Unfortunately, the average deviation does not reflect the relative enhancing and weakening effects of large and small deviations in a set of measurements. This is better achieved by the following measure of precision.

Standard Deviation. This is a root mean square average of

the individual deviations. If $\bar{a}$ is the average deviation and σ is the standard deviation, then the two terms are approximately related by the expression

$$\bar{a} = 0.8\sigma \qquad (2\text{-}7)$$

where

$$\sigma = \frac{d_1^{\,2} + d_2^{\,2} + d_3^{\,2} + \ldots + d_n^{\,2}}{n - 1} \qquad (2\text{-}8)$$

When n is very large, the difference between n and n - 1 is insignificant and the denominator becomes n. When n is small, the statistical distribution over a small number of measurements may be faulty, and this lower degree of precision is reflected in the use of n - 1 in the denominator of Eqn. (2-8). There are two points of interest here: What is the significance of the standard deviation and how can it be used in a decision concerning the reliability of a given result? These points are considered in the next two paragraphs.

Significance Levels. With a normal frequency distribution, we would expect to find that about 1% of the deviations would exceed 2.5σ. This would correspond to what is termed the 99% significance level. About 5% of the deviations would exceed 2.0σ, which would correspond to the 95% significance level. Finally, about 50% of the deviations would exceed 0.67σ, a value which is sometimes called the "probable deviation".

Rejection of Doubtful Measurements. This is a point of some concern, and it is possible to suggest a criterion for the rejection of one measurement which is "obviously" out of line with the other results in a given set. The criterion suggested by Kolthoff and Sandell[1] requires that there be at least and probably not more than ten results and only one doubtful result may be excluded. This criterion suggests that the arithmetic mean ("the average") and the average deviation exclude the doubtful result, and that the doubtful result be rejected if deviation of the doubtful result from the arithmetic mean is more than 2.5 times the average deviation ($\bar{a}$), or 2.0 times the standard deviation. If the deviation of the doubtful result is less than 2.5 times the average deviation, the result should be retained and included in the recalculation of the arithmetic mean.

2.6 Problem Types

Many of the problems involved in marine chemistry can be classified as one of the following types, although some may involve a combination of these types.

Problems involving mass:

1. Percentage composition [see Section 2.7(a)].
2. Chemical formula from elementary composition [Section 2.7(6)].
3. Weight-weight relationships in chemical reactions [Section 2.7(c)].

Problems involving volume:

4. Correction of gas volumes for pressure and temperature changes [Section 2.8(a)].
5. The gram molecular volume concept [Section 2.8(b)].

Problems involving solutions:

6. Problems involving concentrations of solutions [Section 2.9(a)].
7. Problems involving chemical equilibrium [Section 2.9(b)].

Conversions involving length, width, etc. are given in Appendix A.

Before beginning these problems, a review of certain terms may be useful.

Atomic weight represents, properly, the weight of one atom of an element, expressed in atomic mass units (amu). For silicon, the atomic weight is 28.086 amu, where 1 amu is 1/12 the weight of a carbon-12 atom (whose atomic weight is 12 amu).

Molecular weight represents, properly, the weight of one molecule of a compound, expressed in amu. For carbon dioxide (CO_2), the molecular weight is 44.0 amu.

Gram atomic weight (g-at wt) is the weight in grams of Avogadro's number of atoms. For silicon, the gram atomic weight is 28.086 g.

Gram molecular weight (g-mole wt) represents the weight in grams of Avogadro's number of molecules. For carbon dioxide, one gram mole would be 44 g.

These definitions represent certain conversion factors

which are used in the problems that follow. For example, gram molecular weight represents the following relationship:

Since 44 g CO_2 = one mole CO_2, it follows that

$$\frac{1 \text{ mole } CO_2}{44 \text{ g } CO_2} = 1$$

and since one mole $CO_2 = 6.02 \times 10^{23}$ molecules CO_2, it also follows that

$$\frac{6.02 \times 10^{23} \text{ molecules}}{1 \text{ mole } CO_2} = 1$$

$$\frac{1 \text{ mole } CO_2}{6.02 \times 10^{23}} = 1$$

Thus, the number of moles of CO_2 represented by 100 g of Dry Ice would be

$$100 \text{ g } CO_2 \times \frac{1 \text{ mole } CO_2}{44 \text{ g } CO_2} = 2.27 \text{ moles } CO_2$$

Here, the first relationship represents the appropriate conversion factor because the units cancel and the appropriate units of moles of CO_2 are obtained. Any other factor would have resulted in the wrong answer as indicated by incorrect units. This approach, the factor dimensional approach, will be used in the model examples that follow.

2.7 Model Examples - Mass

(a) Percentage Composition

A chemical formula represents the composition of a substance as well as one molecule or one mole. Explicit in the chemical formula is the number and kind of atoms and gram atomic weights of each element. Implicit in the formula is the percentage composition. Although this is an important calculation, it is an uncommon one in marine chemistry and is covered elsewhere.

(b) Chemical Formula from Percentage Composition

Assimilation of nitrogen and phosphorus by marine phytoplankton involves highly selective organic synthesis. It has been of considerable interest to determine the atomic ratios for assimilation of carbon, nitrogen, and phosphorus, which seem to approach 100:16:1, respectively, for many aquatic plants.[2] The atomic ratios are determined from elementary composition data and the following example should illustrate the steps involved.

Model Example. The per cent of ash-free dry weight (i.e., per cent organic matter) of carbon, hydrogen, nitrogen, and phosphorus has been summarized for a variety of aquatic plants.[3] For diatoms, the following elementary composition would be in the observed range of analysis: 53.90% carbon, 10.70% nitrogen, 1.47% phosphorus. What is the carbon: nitrogen:phosphorus atomic ratio?

Method of Solution. As a convenient weight basis, let us take 100 g of the substance; then the percentage of each element gives the number of grams of that element in 100 g of that substance. Obtain the number of gram atomic weights of each element in 100 g of the substance by dividing the actual weight by the appropriate gram atomic weight. Thus

$$\text{C:}\quad \frac{53.90\text{ g}}{12.01\text{ g/g-at wt}} = 4.529$$

$$\text{N:}\quad \frac{10.70\text{ g}}{14.01\text{ g/g-at wt}} = 0.7637$$

$$\text{P:}\quad \frac{1.47\text{ g}}{30.97\text{ g/g-at wt}} = 0.04884$$

Next, calculate the <u>relative</u> number of gram atomic weights (which is equivalent to the relative number of atoms) of each element in terms of the constituent present in the smallest amount (in this example, phosphorus).

$$\text{C:}\quad \frac{4.529\text{ g-at wt C}}{0.04884\text{ g-at wt P}} = 96.72 \text{ or } 97\ \frac{\text{g-at wt C}}{\text{g-at wt P}}$$

$$\text{N:}\quad \frac{0.7637\text{ g-at wt N}}{0.04884\text{ g-at wt P}} = 15.64 \text{ or } 16\ \frac{\text{g-at wt N}}{\text{g-at wt P}}$$

$$\text{P:}\quad \frac{0.04884\text{ g-at wt P}}{0.04884\text{ g-at wt P}} = 1.00\ \frac{\text{g-at wt P}}{\text{g-at wt P}}$$

The atomic ratio is thus C:N:P = 97:16:1 or $C_{97}N_{16}P$. (Note 3.)

(c) <u>Weight-Weight Relationships</u>

In order to do this type of calculation, the balanced chemical equation must be available to indicate the weight relationships. Typically, one is concerned with the weight of product that can be obtained from a given weight of reactant, other reactants being present in excess quantities (as in the determination of sulfate as barium sulfate); or the weight of material that reacts with a given weight of reagent (as in the standardization of hydrochloric acid, Chapter 4), or, in general, the weight (volume, or number of moles) of reagent associated with a weight (volume, or number of moles) of a second reagent. The chemical equation is explicit in terms of moles of reagents; therefore, the first step is to convert a weight or volume of a reagent into the number of moles of that reagent. The second step is to use the weight relationship indicated by the chemical equation to obtain the number of moles of desired reagent. The third step is to convert (if necessary) from moles of desired reagent to the appropriate units.

Model Example. Calculate the weight of pure calcium carbonate needed to produce 0.01 mole of carbon dioxide.

Method of Solution. The balanced equation for the reaction

$$CaCO_3 + 2HC\ell \rightarrow CaC\ell_2 + CO_2 + H_2O$$

shows that one gram molecular weight of carbon dioxide results when one mole of calcium carbonate (100 g) is treated with at least two moles of hydrochloric acid. Thus, weight of calcium needed

$$= 0.01 \text{ mole } CO_2 \times \frac{1 \text{ mole } CaCO_3}{1 \text{ mole } CO_2} \times \frac{100 \text{ g } CaCO_3}{1 \text{ mole } CaCO_3}$$

$$= 1.00 \text{ g}$$

2.8 Model Examples - Volume

(a) Correction of Gas Volumes for Pressure and Temperature Changes

It is often desirable to determine the volume occupied by a definite amount of gas at a certain pressure and temperature. This calculation requires only a knowledge of the volume of the same amount of gas at some other pressure and temperature, and two gas laws. The volume of a gas varies inversely with the pressure to which it is subjected (Boyle's Law) and directly with its absolute temperature (Charles's Law). Commonly, gas volumes are calculated for standard conditions (abbreviated as SC or STP): 0°C (or 273°K) and one atmosphere (760 mm of mercury) pressure.

It is also necessary to remember that gases may be wet, as when they are collected over water or are in contact with water. In this case, one must correct for the exerted water vapor pressure by deducting the vapor pressure of water at the temperature used from the total pressure of the combined gases (Dalton's Law).

Model Example. Calculate the volume that 250 mℓ of carbon dioxide in contact with water at 23°C and under a barometric pressure of 759 mm would have if measured dry at standard conditions.

Method of Solution. First, correct for the vapor pressure of water to convert to a dry basis. The vapor pressure of water at 23°C is 20.9 mm Hg; therefore, the pressure exerted by the dry gas is 738 mm Hg (759 - 21 mm Hg). In changing from this pressure to the greater standard pressure, the volume of the gas would decrease. The volume must be multiplied by a pressure correction factor which is less than unity, or 738/760 (smaller number on top). Also, a decrease in temperature from the initial value of 296°K (273 + 23) to standard temperature of 273°K would be accompanied by a decrease in volume. To achieve this, the initial dry volume is multiplied by a temperature correction which is less than unity, or 273/296 (smaller number on top). Thus

volume at SC

= dry volume x pressure correction x temperature correction

$$\text{volume at SC} = 250\ \text{m}\ell \times \frac{738}{760} \times \frac{273}{373} = 241\ \text{m}\ell$$

(b) The Gram Molecular Volume Concept

The gram molecular volume of a gas is the volume occupied by one mole of that gas. This volume is the same, or nearly so, for all gases under the same conditions of temperature and pressure. At standard conditions, one gram molecular volume is 22.4 liters. The mole (or weight)-volume relationship may be used[4] in determining (a) weight-volume relationships in chemical reactions, (b) molecular weights of gases and volatile liquids (c) weights of given volumes of gases, and (d) relative densities of various gases.

Model Examples. Calculate the weight of calcium carbonate needed to produce 250 mℓ of carbon dioxide at 23°C and a pressure of 758 mm Hg.

Method of Solution. This information is needed to calibrate the apparatus used in the gasometric determination of carbonate (Chapter 33). The solution can be divided into two parts: (a) the calculation of the number of moles of carbon dioxide represented and (b) the weight of calcium carbonate needed.

The first part of the problem may be solved using the gram-molecular volume concept (correcting volume to standard conditions and determining what fraction of a mole or 22.4 liters is represented). Alternatively, the ideal gas law may be used:

$$PV = nRT \tag{2-9}$$

where P represents the pressure (mm Hg), V the volume (mℓ), n the number of moles of gas present, T the absolute temperature, and R is a constant (62,400 mm Hg mℓ $mole^{-1}$ deg^{-1}). Thus, the moles of carbon dioxide present is given by

$$n = \frac{PV}{RT} = \frac{(758 \text{ mm Hg})(250 \text{ m}\ell)}{(62{,}400 \text{ mm Hg m}\ell \text{ mole}^{-1}\text{deg}^{-1})(296°\text{K})}$$

$$= 0.00986 \text{ mole}$$

The second part of the problem has been solved in a previous model example [Section 2.7(c)].

2.9 Model Examples - Solutions

(a) Concentration of Solutions

The concentration units used in marine chemistry include: percentage (by weight or by volume), molarity, and normality (Note 4). Percentage by weight (%w/w) refers to the weight of solute per 100 g of solution; percentage by volume (%v/v) refers to the volume of solute per 100 mℓ of solution. Molarity (M) refers to the number of gram molecular weights per liter of solution. Often, it is convenient to use weight molarity $\underline{M}_w$, which refers to the number of gram equivalent weights per kilogram of sea water. Finally, normality ($\underline{N}$) of a solution refers to the number of gram equivalent weights of solute per liter of solution (Note 4). Normality is primarily useful in titrations because equal volumes of two solutions of the same normality are chemically equivalent.

Model Example. Prepare five liters of 0.1 $\underline{M}$ H_2SO_4 solution from concentrated sulfuric acid 96% (w/w), sp.gr.1.84.

Method of Solution. This is a type of problem that is commonly encountered. The solution depends upon breaking the problem into steps as follows: (a) What is the total number of moles of H_2SO_4 needed (definition of molarity)? (b) What is the total weight of sulfuric acid (definition of gram molecular weight)? (c) What is the weight of concentrated H_2SO_4 needed (definition of weight percent)? (d) What is the volume of sulfuric acid needed (definition of density or specific gravity)? The answers to the individual steps are indicated in the calculation below.

Volume of conc. acid needed (mℓ)

$$= 5 \text{ liters solution} \times \frac{0.1 \text{ mole } H_2SO_4}{\text{liter solution}}^{(a)} \times \frac{96 \text{ g } H_2SO_4}{1 \text{ mole } H_2SO_4}^{(b)}$$

$$\times \frac{100 \text{ g conc } H_2SO_4}{98 \text{ g } H_2SO_4}^{(c)} \times \frac{1 \text{ m}\ell \text{ conc } H_2SO_4}{1.84 \text{ g conc } H_2SO_4}^{(d)}$$

$$= 26.6 \text{ m}\ell$$

Thus five liters of 0.1 $\underline{M}$ H_2SO_4 solution would be prepared by pouring 26.3 mℓ of concentrated sulfuric acid into a large volume of water and diluting to five liters.

Model Example. What is the normality of a 0.1 M sulfuric acid solution?

Method of Solution. Recall the definition of normality and gram equivalent weight (Note 4). One mole of sulfuric acid represents two gram equivalent weights because there are two replaceable hydrogens. Thus, a 0.1 M sulfuric acid solution is also a 0.2 N solution, as seen from the following:

$$\text{normality} = \frac{0.1 \text{ mole } H_2SO_4}{\text{liter solution}} \times \frac{2 \text{ g-equiv wt } H_2SO_4}{1 \text{ mole } H_2SO_4} = 0.2\ \underline{N}$$

(b) Calculations Involving Chemical Equilibria

The expressions for several equilibrium constants have been derived in Chapter 3. These expressions are more conveniently used in the logarithmic form when dealing with marine equilibria because linear relationships are obtained. For example, consider the expression for K_A for a weak acid. The general expression [Eqn. (3-7), Chapter 3] may be used in the logarithmic form:

$$\log K_A = \log(H^+) + \log(A^-) - \log(HA) \qquad (2\text{-}10)$$

Multiplying both sides of the equation by -1 and recalling [Eqn. (2-1)] the definition of p(x), the following expression is obtained:

$$pK_A = pH + p(A^-) - p(HA) \qquad (2\text{-}11)$$

As Sillén has noted,[5] this logarithmic form lends itself to graphical representation as master variable plots. For example, in acid-base systems for which the total concentration is known, the controlling or master variable is the hydrogen ion activity or pH. The concentration of each species in the acid-base system is a unique function of the master variable pH. For the halide ion-silver ion system, the master variable is the concentration of silver ion, or pAg [Eqn. (2-3)]. Other master variables include pE (Note 5) for redox equilibria and log P_{O_2} for equilibria involving oxides and gaseous oxygen.

Admittedly, the master variable plots yield the same information that is obtained from numerical calculations. On the other hand, the relationships between pertinent species and the equivalence points become more readily apparent, particularly with practice.

As an example of a master variable plot, the pH diagram for sea water is represented by Figure 2-1. Here, the concentrations of major acids and bases are plotted as a function of pH. The total concentrations and K_A values used are given elsewhere (Note 6). The important species are: hydrogen and hydroxide ion; boric acid (HB) and borate ion (B^-); and carbonic acid (H_2C), bicarbonate ion (HC^-), and carbonate ion (C^{2-}). The arrows point to three pH values that are of especial interest: the pH of sea water (about 8); and P_1 and P_2, the two equivalence points in the alkalinity titration (Chapter 4). The first equivalence point, P_1, corresponds to

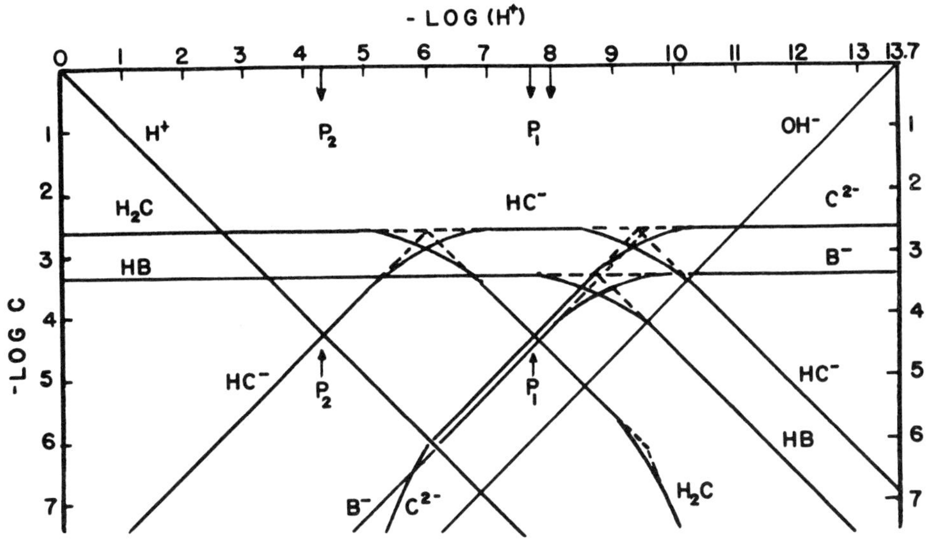

FIG. 2-1

Master variable pH diagram for sea water (redrawn, with permission of American Chemical Society, from Sillén,[5] p. 47).

a conversion of all of the borate ion to boric acid and all of the carbonate to bicarbonate; the second equivalence point, P_2, corresponds to conversion of all of the borate ion to boric acid and all of the bicarbonate to carbonic acid.

The method of calculation of data for the master variable plot is illustrated in the following example.

Model Example. The total boron concentration in sea water is equal to $10^{-3.63}$ and the pK_A value for boric acid $B(OH)_3$ is 8.8. Calculate the concentration of boric (HB) acid and borate (B^-) ion as a function of pH.

Method of Solution. There are three regions of interest:

at very high pH, nearly all of the boron is in the form of borate ion; at very low pH, nearly all the boron is in the form of boric acid; at an intermediate region (pH=7.5-9.3), the fraction of each form is significant.

For the first pH region (pH=11-14), let x equal the concentration of boric acid. The concentration of borate ion would be the total concentration of boron minus the amount that exists as boric acid:

$$(B^-) = 10^{-3.63} - x = 10^{-3.63} \qquad (2\text{-}12)$$

This assumption is valid as long as x is less than 10% of $10^{-3.63}$; therefore $-\log(B^-) = -3.63$, or $p(B^-) = 3.63$.

Rearranging the relationship derived earlier [Eqn. (2-10)] calculate the value of -log(HB) and substitute appropriate values of pH:

$$-\log(HB) = -pK_A + pH + p(B^-) \qquad (2\text{-}13)$$

at pH = 14, $-\log(HB) = -8.8 + 14 + 3.63 = 8.8$; $\log(HB) = -8.8$

at pH = 12, $-\log(HB) = -8.8 + 12 + 3.63 = 6.8$; $\log(HB) = -6.8$

at pH = 10, $-\log(HB) = -8.8 + 10 + 3.63 = 4.8$; $\log(HB) = -4.8$

Thus for each pH selected, values are obtained for $\log(B^-)$ and log(HB). At a pH of less than 9.5, the magnitude of x increases to the point that the assumption made earlier [Eqn. 2-11)] is not valid.

For the second pH region (pH=1-7) let y be the concentration of borate ion. Then,

$$(HB) = 10^{-3.63} - y = 10^{-3.63} \qquad (2\text{-}14)$$

This assumption is made on the same basis as before [Eqn. (2-12)], and log(HB) = -3.63, or p(HB) = 3.63. The method of calculation is essentially the same. The relationship used earlier [Eqn. (2-13)] is rearranged for convenience as follows:

$$-\log(B^-) = pK_A - pH + p(HB) \qquad (2\text{-}15)$$

at pH = 5, $-\log(B^-) = 8.8 - 5 + 3.63 = 7.4$; $\log(B^-) = -7.4$

at pH = 6, $-\log(B^-) = 8.8 - 6 + 3.63 = 6.4$; $\log(B^-) = -6.4$

at pH = 7, $-\log(B^-) = 8.8 - 7 + 3.63 = 5.4$; $\log(B^-) = -5.4$

For the third pH region (pH=7.5-9.3), let y equal the concentration of borate ion. The concentration of boric acid would be $(10^{-3.63} - y)$ and y is not negligible with respect to $10^{-3.63}$. Here, the conventional equilibrium expression becomes useful:

$$K_A = \frac{(H^+)(B^-)}{(HB)} = 10^{-8.8} = 1.8 \times 10^{-9} \qquad (2\text{-}16)$$

At a pH of 9, the appropriate values may be substituted and the following expression solved for y:

$$1.8 \times 10^{-9} = \frac{(1 \times 10^{-9})(y)}{(2.34 \times 10^{-4} - y)} \qquad (2\text{-}17)$$

and we have

$$4.2 \times 10^{-13} - 1.8 \times 10^{-9}y = 1 \times 10^{-9}y$$

$$2.8 \times 10^{-9}y = 4.20 \times 10^{-13}$$

$$(B^-) = y = \frac{4.20 \times 10^{-13}}{2.8 \times 10^{-9}} = 1.5 \times 10^{-4}$$

$$= 10^{-3.82}$$

$$\log(B^-) = -3.82$$

$$(HB) = 2.34 \times 10^{-4} - y = 2.34 \times 10^{-4} - 1.4 \times 10^{-4}$$

$$= 1.84 \times 10^{-4} = 10^{-3.74}$$

$$\log(HB) = -3.74$$

NOTES

1. Portions of the material in Sections 2.2-2.5 are taken from Laboratory Chemistry, T. Moeller and D. F. Martin, 1965. Reproduced by permission of D. C. Heath and Company, Boston.
2. Natural or Napierian logarithms use a base e; $\log_e 10 = 2.30259$.
3. This is the method for obtaining the simplest or empirical formula. If the molecular weight is known, the true formula can be found because the true formula is a multiple of the empirical formula.
4. A gram equivalent weight is that weight (in grams) chemically equivalent to 1.00797 g of hydrogen (0.5 mole). It may be useful to note the utility of normality as a concentration unit lies in the relationship:

$$N_1 \times m\ell_1 = N_2 \times m\ell_2$$

and the mg-equiv wt of solute = 10^{-3} equiv wt of solute = $N_1 \times m\ell_1$. Here, the subscripts refer to the normality and volume of two reacting species in a titration.
5. The quantity pE is a measure of the electron activity in

solution; $pE = -\log(e^-)$. The quantity pE is obtained by dividing the oxidation potential by the factor 0.05915v at 25°C ($=RTF^{-1}$), where the symbols are those defined in Chapter 4.

6. Values used[5] include: total boron concentration, $10^{-3.37}$ M; total inorganic carbon, $10^{-2.62}$; and the following dissociation constant values: H_2C, 6.0; HC^-, 8.8; and pK_w, 13.7.

REFERENCES

1. I. M. Kolthoff and E. B. Sandell, Textbook of Quantitative Inorganic Analysis, 3rd ed., Macmillan, New York, 1952, p. 276.

2. Cf. R. F. Vaccaro, in Chemical Oceanography, (J. P. Riley and G. Skirrow, eds.) Academic Press, New York, 1965, p. 374.

3. J. H. Ryther, Limnol. Oceanog., 1, 72 (1956).

4. T. Moeller and D. F. Martin, Laboratory Chemistry, Heath, Boston, 1965, pp. 241-244.

5. L. G. Sillén, Adv. Chem. Ser., 67, 45-56 (1967).

3

ACIDS, BASES, AND pH OF NATURAL WATERS

3.1 Introduction

One of the prominent characteristics of natural waters is the presence and concentration of acids and bases. The composition of these materials is governed by physical and chemical processes. Physical processes include removal of water by evaporation or freezing or addition of water by precipitation and melting. Chemical processes include addition or removal of carbon dioxide by air-water interaction, precipitation or dissolution of calcium carbonate, ion-exchange processes between natural waters, and suspended materials or sediments.

A subsequent determination (Chapter 4) is concerned with finding the concentration of certain bases in sea water. However, at this point, it seems useful to review the modern terminology of acids, bases, and the related concept, pH.

There are a variety of definitions of acids and bases, including the definitions of Arrhenius, Brønsted-Lowry, Lewis, Usanovich, and Pearson.[1-4] Of these, the Brønsted-Lowry system is the most useful for our purpose.

According to the Brønsted-Lowry concept, an acid is a

species that can donate a proton to another species; a base is a species capable of accepting a proton. This concept implies the existence of an acid-base pair in water. Consider a weak acid HA which is dissolved in water and is represented by the equation

$$\underset{\text{acid}_1}{HA} + \underset{\text{base}_2}{H_2O} \rightleftarrows \underset{\text{acid}_2}{H_3O^+} + \underset{\text{base}_1}{A^-} \qquad (3\text{-}1)$$

The species HA and A^- are tied together or conjugated by the loss and gain of a proton. The acid HA is said to be the conjugate acid of the base A^-; A^- is the conjugate base of the acid HA.

Or, consider base B which is dissolved in water:

$$B + H_2O \rightleftarrows BH^+ + OH^- \qquad (3\text{-}2)$$

If B is NH_3, the conjugate acid is ammonium ion. When Equations (3-1) and (3-2) are compared, it is apparent that water is amphiprotic, that is, it can behave both as an acid and as a base.

The acids that would be encountered in sea water are of two types: oxyacids (carbonic, dihydrogenphosphate, boric, and silicic acids) and hydrated metal ions, $M(H_2O)_x{}^{n+}$ (Note 1):

$$M(H_2O)_x{}^{n+} + (H_2O) \rightleftarrows M(OH)\frac{(n-1)}{x-1} + H_3O^+$$

All are weak acids, and those of the second type are present in too small a quantity to make any significant contributions to the acidity. Any anion is a Brønsted-Lowry base, and in sea water, the major anions are C^-, Br^-, $SO_4{}^{2-}$, $HCO_3{}^-$, $CO_3{}^{2-}$, and

$H_2BO_3^-$. Of these, only the last three make a significant contribution to the alkalinity of sea water through the reactions:

$$HCO_3^- + H_2O \rightleftarrows H_2CO_3 + OH^- \qquad (3\text{-}3)$$

$$CO_3^{2-} + H_2O \rightleftarrows HCO_3^- + OH^- \qquad (3\text{-}4)$$

$$H_2BO_3^- + H_2O \rightleftarrows H_3BO_3 + OH^- \qquad (3\text{-}5)$$

One measure of the strength of an acid or a base is the magnitude of the equilibrium constant K:

$$K = \frac{(H_3O^+)(A^-)}{(HA)(H_2O)} \qquad (3\text{-}6)$$

where the parentheses represent activities (Note 2) of the species. Because the activity of water is constant,

$$K_A = \frac{(H_3O^+)(A^-)}{(HA)} \qquad (3\text{-}7)$$

where K_A is the conventional acid dissociation constant. A similar expression may be written for the base dissociation constant K_B:

$$K_B = \frac{(BH^+)(OH^-)}{(B)} \qquad (3\text{-}8)$$

From the foregoing discussion, it is apparent that the concentration of various bases controls the alkalinity of natural waters. It is equally true, but perhaps less apparent, that the alkalinity, or better, the hydrogen ion concentration, controls the concentration of various acids and bases. Thus the hydrogen ion concentration or the pH is a <u>master variable</u> that formally governs the concentration of other species. The usefulness of this view will become apparent later.

First, it is necessary to note that the value of the equilibrium constant is dependent upon the medium or activity scale (and, of course, the temperature) in which it is measured. Second, it is appropriate to consider what we mean by the term pH. After we consider these points in succeeding sections, we can consider some applications of equilibrium constants in marine chemistry problems (Chapter 2).

3.2 Activity Scales[5,6]

Two activity scales have been used in treating ionic equilibria in solution: the infinite dilution activity scale and the ionic medium method. The former is useful in considering equilibria in fresh water, the latter in considering equilibria in sea water.

The Infinite Dilution Activity Scale. This is the traditional scale and is so defined that the activity coefficient γ_x [which is the ratio of the activity of a species and the concentration of that species, Eqn. (3-9)], approaches

$$\gamma_x = \frac{(x)}{[x]} \tag{3-9}$$

unity as the solution approaches infinite dilution. Actually, the value of γ_x is approximately unity as long as the concentrations of ionic species are low, typically less than 10^{-3} M.

The Ionic Medium Activity Scale. This scale is so defined that the activity coefficient γ_x approaches unity as the solution approaches the concentration of the pure medium.

The medium is a concentrated salt solution, e.g., 2 M sodium chloride. Sea water is an example of an ionic medium of nearly constant composition.

On the ionic medium activity scale, the equilibrium constant is defined as the limiting value (in pure medium) for the concentration quotient Q,

$$K = Q_{lim} \tag{3-10}$$

On the infinite dilution scale, the two constants are related by the expression

$$K_A = \frac{[H_3O^+][A^-]}{[HA]} \frac{H_3O^+ \ \gamma_{A^-}}{\gamma_{HA}} \tag{3-11}$$

$$K_A = Q_A \frac{\gamma_{H_3O^+} \ \gamma_{A^-}}{\gamma_{HA}} \tag{3-12}$$

At low concentrations of reactants, values of K and Q are equal, within experimental error.

3.3 pH

A vast number of pH measurements of sea water have been recorded, and it seems appropriate to consider the validity and usefulness of these measurements.

In the past, pH has been defined in at least three different ways (Note 3).

Sorensen's Definition. According to the original definition, $pH = -\log[H^+]$. This definition may be used with the ionic medium scale because it is possible to measure accurately the relationship $-\log[H^+] = -\log(H^+)$.

Activity Definition. On the "infinite dilution" scale, $pH = -\log(H^+) = -\log \gamma_H - \log[H^+]$. This definition is inapplicable at the ionic strength (Note 4) of sea water because the estimate of γ_H becomes unreliable as the ionic strength increases from 0.01 to 0.2.

Operational Definition. Several workers[7,8] have suggested the use of an operational definition by which pH is defined by a standard cell and the pH of the unknown solution(pH_x) is determined relative to the pH of a standard buffer solution (pH_s). The value of pH_s is estimated in terms of the activity definition.

One standard cell would be calomel electrode |solution| glass electrode, and, in two separate experiments, the emf values are determined for the unknown solution (E_x) and the standard buffer solution (E_s), which has a pH of pH_s. With this information, the pH of the unknown solution is calculated from the relationship (Note 3):

$$pH_x - pH_s = -(E_x - E_s)(RTF^{-1} \ln 10)^{-1} \quad (3\text{-}13)$$

The chief advantage of the operational definition method is that a reproducible number always results. The chief disadvantage is that pH contains two uncertainties, one in the definition of γ_H and a second in the unknown liquid junction potential (Note 5) which may be significant for sea water. Nevertheless, the pH values obtained from emf measurements are comparable within 0.01 for sea water at the same salinity, temperature, and depth.

There is reason to question the usefulness of pH measurements because pH varies when the water sample is brought to a different temperature and pressure, and comparison of pH values at the different conditions is inherently difficult.

The problem of comparison may be overcome if the total carbonate concentration C_t and the total alkalinity A_t are measured, because these two values are conservative and independent of pressure and temperature. These quantities may be determined by a single emf titration which is quite accurate and which is described in the next chapter. As an introduction to this chapter, directions for the determination of pH are given in the following section.

3.4 Measurement of pH

Procedure[8]

(a) Apparatus

Electrodes. For practical use, a glass electrode is used in conjunction with a calomel reference electrode. More recently, a combination electrode has become available, and these are recommended for practical use. The glass electrode or combination (glass-calomel) electrode consists of a glass cylinder terminating in a thin glass bulb, which can be easily scratched or cracked. In either case, the usefulness of the electrode can be impaired or destroyed. The electrode should be stored in distilled water between uses and must be soaked overnight before being used either for the first time or after having been stored. Other precautions and directions for use are supplied with the individual electrode.

pH Meter. A variety of pH meters are available. As a minimum requirement, it should be possible to read the pH scale to within ±0.02 pH units.

(b) Reagents

Buffer solutions may be purchased from the manufacturer of the pH meter, or may be prepared according to the directions that follow. In either case, the buffer solutions used for standardization should be selected so that their pH is close to that of the sample. Preferably, two buffer solutions should be selected to bracket the pH of the sample. The solutions should be prepared with freshly boiled distilled water (Chapter 9) and be protected from carbon dioxide. Some workers add a small crystal of thymol as a preservative.

pH 1.68 Buffer. Dissolve 6.35 g of reagent-grade potassium tetroxalate [$KH_3(C_2O_4)_2 \cdot 2H_2O$] in boiled distilled water (Chapter 9) and dilute to 500 mℓ (Note 6).

pH 3.56 Buffer. Prepare a saturated solution of potassium hydrogen tartrate ($KHC_4H_4O_6$) by suspending 20 g of salt in 400 mℓ of distilled water. The solution should be shaken vigorously, then filtered just before use (Note 7).

pH 4.01 Buffer. Dry reagent-grade potassium hydrogen phthalate at 110° overnight and cool in a desiccator.

Dissolve 15.106 g of dry $KHC_8H_4O_4$ in boiled distilled water and dilute to 500 mℓ (Note 7).

pH 6.86 Buffer. Dry reagent-grade anhydrous potassium dihydrogen phosphate (KH_2PO_4) in an oven at 110° and reagent-grade disodium hydrogen phosphate (Na_2HPO_4) in an oven at 130° overnight; cool in a desiccator. Dissolve 1.720 g of the potassium salt and 1.775 g of the sodium salt in distilled water and dilute to 500 mℓ (Note 6).

pH 9.18 Buffer. Dissolve 1.907 g of reagent-grade sodium borate ($Na_2B_4O_7 \cdot 10\ H_2O$) in distilled water and dilute to 500 mℓ (Note 6).

(c) Standardization of the pH Assembly

Allow the pH meter to warm up, and bring it to electrical balance following the instructions in the instrument manual. Wash the combination electrode thoroughly with distilled water, then with the buffer solution. Place the electrode in buffer solution in a suitable container and allow the electrode to come to the temperature of the solution. Note the temperature and adjust the temperature compensation dial on the pH meter. Adjust the standardizing knob until the pH meter is balanced at the known pH of the buffer solution (pH_s) [Eqn. (3-13)]. It should be possible to replace the solution with fresh buffer solution at the same temperature and, without touching the adjusting knob, obtain the same value of pH (±0.02 pH unit).

Remove the combination electrode and wash it with

distilled water, then with a second standard buffer solution which is at the same temperature as the first. The pH meter and combination electrode is functioning properly if the observed pH value of the second standard agrees with the reported value within ±0.04 units. A recheck of the buffer standard may be made during a series of pH measurements and a final check should be made at the end of the series of measurements.

(d) Determination of pH of Unknown

Once the assembly is standardized, the pH of the unknown sample may be determined. Unfortunately, the pressure and temperature of the sample are usually different from the ambient conditions under which it is measured. For this reason appropriate corrections must be made. Some workers depend upon a rapidity of measurement to minimize the effects of sample constant with the atmosphere; other workers use a Dewar flask to keep the temperature of the sample constant. In either case, it is best to keep the sample from being unduly exposed to the atmosphere by using a closed container (Note 8).

The temperature of the buffer standards should approximate that of the unknown sample; otherwise, the value of pH_s should be corrected for temperature changes. Measure the pH of the unknown sample by washing the electrode with distilled water, then with sample. Place the electrode in the sample of the unknown, and allow enough time for the electrode to reach the temperature of the sample. Record the observed pH [pH_x, Eqn. (3-13)]. Repeat the measurement with a fresh sample.

Two successive measurements of the same sample should not differ by more than ±0.02 units (Note 9).

A correction must be made to convert the pH measured under ambient conditions to the pH *in situ*[9]:

$$\text{pH } \underline{\text{in}}\ \underline{\text{situ}} = pH_x + x(t - t') + \frac{\Sigma D}{1000} + \delta\pi_{t^o}(t' - t^o)$$

where t = temperature *in situ*

t' = temperature during measurement

t^o = temperature of the buffer standard

x = a temperature coefficient of pH and chlorinity

Σ = a function of surface pH and depth

D = depth, meters

$\delta\pi_{t^o}$ = a correction factor which must be determined for each instrument.

Additional information, including the values of the corrections that must be made, has been summarized elsewhere.[8-10] These data are not given here because many workers agree with Sillén[5] that total alkalinity and total carbonate are more useful measurements.

NOTES

1. The metal[1] is a transition metal ion such as copper or nickel and is not sodium, potassium, calcium, or magnesium.

2. The activity of a species is a measure of its <u>effective</u> concentration, as opposed to the total concentration (indicated by bracketed species). The activity of a species is related to the concentration by an activity coefficient γ_x; $(X) = \gamma_x[X]$. More rigorously, the activity of a species (X) is defined by the relationship $\mu_x = \mu_x^* + RT\ \ell n(X) = \mu_x^* + RT\ \ell n\ \gamma x\ [X]$. Here, μ_x is the chemical potential; μ_x^* is a constant that defines the activity scale; γx is the activity coefficient; and the other terms have the usual meaning (see Chapter 4).

3. The material in this and the preceding section was adapted from Sillén,[5,6] and additional references may be found elsewhere.[8]

4. Ionic strength is a measure of the number and kind of ions in a solution. It is defined as being equal to $\Sigma\ c_i z_i^2$ where c_i is the concentration of an ion and z_i is the charge on that ion.

5. There is a liquid junction at the interface of two solutions of the cell. Here, one solution is unknown with pH_x and the second is the potassium chloride solution in the calomel electrode. It is not possible to measure directly the liquid junction's potential or the emf across the interface, but it is possible to minimize the effects of this potential.

6. The reported pH values of the standard buffer refer to 25°C. The following temperature variations are reported[8]: pH 1.68 buffer: 0°, 5°, 10°, 15°: 1.67; 20°, 25°: 1.68; 30°, 35°: 1.69; 40°: 1.70. pH 3.56 buffer: 25°, 3.25; 30°, 35°: 3.55; 40°, 3.54. pH 4.01 buffer: 0°, 5°; 4.01; 10°, 15°, 20°: 4.00; 25°, 30°: 4.01; 35°: 4.02; 40°: 4.03. pH 6.86 buffer: 0°; 6.98; 5°: 6.95; 10°: 6.92; 15°: 6.90; 20°: 6.88; 25°: 6.86; 30°: 6.85; 35°, 40°: 6.84. pH 9.18 buffer: 0°, 9.46; 5°, 9.39; 10°, 9.33; 15°, 9.33: 20°, 9.27; 25°, 9.18; 30°, 9.14; 35°, 9.10; 40°, 9.07.
7. The pH of this buffer is rather insensitive to change with concentration. A 10% change in concentration produces a change in pH of 0.01.
8. The buffer and test solutions should be tested in a container made from a weighing bottle and fitted with a three-hole rubber stopper. The stopper should have holes drilled to accomodate the combination electrode, thermometer, and drying tube filled with Ascarite.
9. If the seawater sample is in contact with atmosphere, a drift of 0.05-0.10 pH units may be observed.

REFERENCES

1. W. Luder and S. Zuffanti, The Electronic Theory of Acids and Bases, Wiley, New York, 1946.

2. R. P. Bell, Acids and Bases, Methuen, London, 1952.

3. T. Moeller, Inorganic Chemistry, Wiley, New York, 1952.

4. R. G. Pearson, Science, 151, 172 (1966).

5. L. G. Sillén, Adv. Chem. Ser., 67, 45-56 (1967).

6. L. G. Sillén, Am. Assoc. Advan. Sci. Publ., 67, 549 (1961).

7. D. A. MacInnes, Science, 108, 693 (1948).

8. R. G. Bates, Determination of pH, Wiley, New York, 1964.

9. K. Buch and S. Gripenberg, J. Cons. int. Explor. Mer, 7, 242 (1932).

10. K. Buch and O. Nynas, Acta Acad. åbo., 12, 26 (1939).

4

ALKALINITY AND TOTAL CARBONATE

4.1 Introduction

Sea water contains several protolytic species, including carbonate [~2.4 mM_w (millimole/kg of sea water)], borate (~0.43 mM_w), phosphate (~0.0023 mM_w), silicate, and fluoride, in various stages of protonation. In terms of contribution to alkalinity, most species except inorganic carbon and boron can be neglected (Note 1).[1-6]

The concentrations of protolytic species are characterized by three parameters: total alkalinity A_t, total carbonate concentration C_t, and total borate concentration B_t:

$$A_t = C_{HCO_3^-} + 2C_{CO_3^=} + C_{H_2BO_3^-} + (C_{OH^-} - C_{H^+}) \quad (4\text{-}1)$$

$$C_t = C_{H_2C} + C_{HCO_3^-} + C_{CO_3^=} \quad (4\text{-}2)$$

$$= A_t - C_{H_2BO_3^-} - (C_{OH^-} - C_{H^+}) \quad (4\text{-}2a)$$

$$B_t = C_{H_3BO_3} + C_{H_2BO_3^-} \quad (4\text{-}3)$$

(Here, $C_{H_2C} = C_{H_2CO_3} + C_{CO_2}$; see Chapter 8, Volume 2).

The influence of variation in salinity can be eliminated for ease of comparison by using specific alkalinity or normalized alkalinity. Specific alkalinity has values in the range 0.11-0.13, though a value of 0.123 is thought to be "typical"

$$\text{specific alkalinity} = \frac{A_t \times 1000}{\text{chlorinity}} \quad (4\text{-}4)$$

of most seawater samples. The parameter has the disadvantage of "mixed" units, a problem that is obviated if normalized alkalinity values

$$A_n = A_t \times 19.274/(C\ell^{o}/oo) \quad (4\text{-}5)$$

are used. Normalized carbonate alkalinity C_n can be defined similarly. In standard sea water (since $C\ell^{o}/oo = 19.374$), $A_n = A_t$ and $C_n = C_t$.

The normalized values of total alkalinity and carbonate alkalinity can be used to correlate the effect of chemical changes (Table 4-1). In Park's view,[6] the combination of pH and P_{CO_2} (the partial pressure of carbon dioxide in a given mass of sea water) is currently the most precise way to study changes due to biological activity and carbonate dissolution or precipitation. The important consideration is the precision of the various methods, rather than accuracy (Chapter 8, Volume 2).

In the past, the precision has been limited by the methodology. The total alkalinity may be determined by adding an excess of standard acid, boiling off all carbon dioxide, and titrating back to a pH of 6 (methyl red or methyl-orange end-point).[1] Thus all carbonate and bicarbonate has been converted to carbon dioxide and expelled, and all borate has been converted to boric acid. The amount of acid then corresponds to the total alkalinity, Eqn. (4-1). Other methods have been used

TABLE 4-1

Effect of Chemical Changes on Alkalinity and Total Carbonate

Process	ΔA_n	ΔC_n	$\Delta(A_n - C_n)$	$\Delta(A_n - 2C_n)$
Carbonate dissolution and precipitation				
Solution of a moles of $CaCO_3$	+2a	+a	+a	—
Precipitation of a moles	-2a	-a	-a	—
Biological activity				
Addition of b moles of CO_2	—	+b	-b	-2b
Loss of b moles	—	-b	+b	+2b
Ion-exchange				
Exchange of c moles of H^+ for equivalent amounts of Na^+, K^+, $\frac{1}{2}Mg^{2+}$ from clays	+c	—	+c	+c

and reported by Anderson and Robertson,[2] Bruneau, Jerlow, and Koczy,[3] and Tsurikova,[4] among others. A useful summary has been provided.[5]

Also, in the past, most methods have used an indicator to detect the equivalence point(s), with the usual limitation on precision and accuracy. Now, the pH of a seawater-acid sample would be measured by determining the E or emf between glass and reference electrodes in the sample.

From a plot of E (as mV) or pH as a function of volume of standard acid, it may or may not be apparent that there are two equivalence points, P_1 and P_2. At the first point, P_1, the solution contains only HCO_3^- and $H_2BO_3^-$; at the second point, P_2, the solution contains only H_2CO_3 and H_3BO_3(i.e., $[H^+] = [HCO_3^-]$, assuming the concentration of all other protolytic species negligible (see Fig. 2-1). Let v_1 denote the volume of standard acid needed to reach the first equivalence point P_1; and let v_2 denote the volume of standard acid needed to reach the second equivalence point P_2. The value of v_2 corresponds to the total alkalinity A_t, and the value of v_1 corresponds to the total carbonate concentration C_t. Dyrssen has shown[7,8] that the two equivalence points are more obvious when a Gran plot[9] is used.

4.2 Dyrssen Method for Alkalinity and Total Carbonate

The procedure used here involves an emf titration of sea water along with the use of Gran's graphical method[9] for evaluating the equivalent points. The emf (or pH) between a glass electrode and a calomel electrode immersed in a seawater sample is measured by means of a suitable potentiometer after increments of standard hydrochlorid acid. The hydrochloric acid should be 0.1 M with respect to hydrogen ion and should have about the same ionic strength as the sea water being titrated.

From a single titration, it is possible to calculate the total alkalinity and the total carbonate concentration. These quantities may be made independent of pressure and temperature of the seawater sample as collected and titrated *if* the quantities are expressed in the proper units, mM_w.

Procedure[7,8]

(a) Reagents

Standard Hydrochloric Acid Solution. The solution should be 0.100 N and of approximately the same salinity as the sea water being titrated. Prepare the solution from ampoules of concentrated volumetric standardized hydrochloric acid and dilute to one liter with artificial sea water.

An alternative method is to dissolve 8.4 mℓ of concentrated hydrochloric acid in artificial sea water, dilute to one liter, and standardize with standard base (Note 2).

Artificial Sea Water. This solution should contain in 1 kg of solution, 0.478 mole of sodium, 0.064 mole of magnesium, 0.550 mole of chloride, and 0.028 mole of sulfate. Dissolve

24.7 g of sodium chloride, 13.0 g of $MgC\ell_2 \cdot 6H_2O$, and 9.0 g of $Na_2SO_4 \cdot 10\ H_2O$ in 954 mℓ of distilled water.

(b) Analysis

Weigh, or accurately measure with a calibrated pipet, a 200-mℓ sample of sea water (Note 3) in a 250-mℓ beaker equipped with a magnetic stirring bar. Place glass and calomel electrodes, attached to a suitable potentiometer or pH meter, in the beaker (Notes 4,5). The electrodes do not need to be standardized for the purposes of this determination.[7] Measure and record the emf (E or the pH meter reading B) value for the initial sample and after each addition of standard hydrochloric acid (0.20-mℓ increments). Allow time for the emf values to become constant. The acid should be added from a constant-rate buret, an automatic pipet, or an analytical buret. Record the data as indicated in Table 4-2. Add increments of acid until the difference in emf between two consecutive readings is less than 1 mV. Record the temperature of the sample at intervals during the titration, and use a mean value for the calculations. After the titration, rinse the electrodes.

The calculation of A_t and C_t depends upon whether emf or pH data were recorded, and the two possibilities are considered separately in the following sections.

(c) Calculations Using E(mV)

For this system, the relationship between E and (H^+) is

$$E = E^o + a \log(H^+) \quad \text{or} \quad (H^+) = 10^{(E - E^o)/a} \qquad (4\text{-}6)$$

where $a = RTF^{-1} \log 10$ (=59.16 mV at 25°C)

TABLE 4-2

Data for Sample Graphical Calculation of Total Alkalinity

v (mℓ)[a]	E (mV)	F_2	F_1
0	73	—	—
0.200	99	—	—
0.400	120	7.30	—
0.60	135	12.3	—
0.80	145	17.2	—
1.00	150	19.4	—
1.20	158	24.6	—
1.40	164	28.4	—
1.60	170	32.8	—
1.80	174	35.1	—
2.00	177	35.2	—
.	.	.	
.	.	.	—
.	.	.	
3.6	243	—	5.4
3.80	272	—	16.6
4.20	312	—	77.7
4.40	323	—	121
4.60	331	—	163
4.80	335	—	191
5.00	339	—	224
5.20	346	—	296
5.40	349	—	332
5.60	352	—	372
5.80	355	—	417
6.00	357	—	458

[a]Volume of 0.1000 N hydrochloric acid

$R = 8.314 \times 10^{-3}$ mV C deg^{-1} mole^{-1}

$F = 96,520$ C mole^{-1}

T = average absolute temperature during the titration (= °C + 273)

log 10 = 2.303

The following steps are involved in the calculation of two functions for the Gran plot (see Table 4-2).

1. Calculate the value of the function $F_1 = 10^{(E + E_1)/a}$ (where E_1 is an arbitrary constant) for each point in the titration starting from the last and continuing backward.

2. Plot the values of F_1 as a function of milliliters of standard acid v, and obtain a straight line which deviates from linearity as F_1 approaches zero. Draw the best straight line through these points and determine the intercept on the horizontal axis. The value of the intercept is v_2 and, as noted, corresponds to the total alkalinity.

3. Obtain the value of v_1 using the function $F_2 = (v_2 - v)10^{(E + E_2)/a}$ (where E_2 is an arbitrary constant). Plot F_2 versus v. The straight line portion may be extrapolated to $F_2 = 0$ to give the intercept $v = v_1$.

4. From this information, calculate the values of C_t and A_t:

$$C_t = \frac{\text{concentration of standard acid x } (v_2 - v_1)}{\text{amount of sea water}} \qquad (4\text{-}7)$$

$$A_t = \frac{\text{concentration of standard acid x } v_2}{\text{amount of sea water}} \qquad (4\text{-}8)$$

5. An average value of the standard cell potential E^o

for the system may be calculated from the linear portion of portion of F_1:

$$E^o = E - a \log(v - F_2) \times \frac{\text{conc standard acid}}{(v + \text{volume of sea water})} \quad (4\text{-}9)$$

(d) Sample Calculation

Using the data recorded in Table 4-2, and the equations for F_1 and F_2 as given in the preceding section, the values of F_1 and F_2 were calculated. In this case,

$$a = 59.1$$

$$E_1 = -200 \text{ (arbitrary)}, \; E_2 = -100 \text{ (arbitrary)}$$

$$v_2 = 3.76 \text{ m}\ell, \; v_1 = 0.05 \text{ m}\ell$$

$$A_t = \frac{0.100 \times 3.76}{0.1519} = 2.48 \text{ mmoles/kg}$$

$$C_t = \frac{0.100}{0.1519} \times (3.76 - 0.05) = 2.44 \text{ mmoles/kg}$$

(e) Calculations Involving pH Meter Readings

When the potentiometer gives direct readings on a pH scale, and when these are sufficiently precise (i.e., 0.002 units), it is convenient to calculate the Gran plot parameters in terms of pH meter readings (= B). Three expressions may be used:

$$F_1 = (v_o + v)10^{-B} \quad (4\text{-}10)$$

$$F_2 = (v_2 - v)10^{-B} \quad (4\text{-}11)$$

$$F_3 = (v - v_1)10^{B} \quad (4\text{-}12)$$

If v is much less than v_o, as is usually the case, $(v_o + v)$ is a constant, and $F_1 = 10^{-B}$.

The values of F_1 [Eqn. (4-10)] are calculated and the value of v_2 is estimated by extrapolation (steps 2-3 in the preceding section). Values of F_2 are calculated and used to obtain the value of v_1 [step 3, but using Eqn. (4-11)]. Plotting F_3 and F_2 will yield two straight lines of opposite slope that should intersect at v_2. Values of A_t and C_t are calculated as before [Eqns. (4-7) and (4-8)].

It should be noted again for emphasis that it is not necessary to standardize the pH meter in order to obtain pH values for the estimation of alkalinity. The value of B is, of course, related to $-\log(H^+)$ by

$$B = pH_o - \log(H^+) \qquad (4\text{-}13)$$

Here, pH_o is a constant that includes the liquid junction potential, the ratio between two activity scales (sea water and dilute aqueous solution), systematic errors in the electrodes, as well as other systematic errors.

4.3 Total Carbon Dioxide Concentration

This parameter, $\Sigma\ CO_2$, is also of interest and can be approximated (Eqn. 4-14). It can be determined from titration

$$\Sigma\ CO_2 = [H_2CO_3] + [CO_2] + [HCO_3-] \qquad (4\text{-}14)$$

data (Eqn. 4-15); N is the normality of the titrant, and v_1 and v_2 have been defined. It is essential that no atmospheric

$$\Sigma\ CO_2 = (v_2 - v_1)\ \underline{N} \qquad (4\text{-}15)$$

exchange with carbon dioxide occur during the processing of the sample. To this end a special titration flask has been designed and described.[10] With this unit, an accuracy of 0.68% at the 95% confidence level is reported.

NOTES

1. The concentrations of other species may need to be considered, e.g., in some estuaries, the concentration of inorganic phosphate may be significant; in anoxic waters, the concentrations of HS^-, ammonia, ammonium ion, and silicate may be significant.
2. Dissolve 1.021 g of potassium hydrogen phthalate in 25 mℓ of distilled water and titrate with standard base to a phenolphthalein endpoint. Fifty milliliters of 0.1 N acid will be needed. The normality of the standard base is equal to the number of grams of phthalate/mℓ acid x 0.20423).
3. The sample should be handled so as to avoid losses of carbon dioxide and calcium carbonate.
4. It may be possible to connect the glass and calomel electrode in either of two ways. If E decreases with addition of hydrochloric acid, the expression used in the calculations is $F_1 = 10^{(E_1 - E)/a}$.
5. It should be possible to read the potentiometer to 0.6 mV (0.01 pH unit) for best results.

REFERENCES

1. H. Barnes, Apparatus and Methods of Oceanography Part 1: Chemical, George Allen Unwin, London, 1959, pp. 200-205.

2. D. H. Anderson and R. J. Robinson, Ind. Eng. Chem. Anal. Ed., 18, 767 (1946).

3. L. Bruneau, N. G. Jerlov, and F. F. Koczy, Report of Swedish Deep Sea Expedition, 1947-1948, 3 (Physics and Chemistry, No. 4), 99 (1953).

4. A. P. Tsurikova, Tr. Gos. Ikeanogr. Inst., 151 (1962); C. A., 60, 303 (1964).

5. G. Skirrow, in Chemical Oceanography, (J. P. Riley and G. Skirrow, eds.), Vol. 1, Academic Press, New York, 1965, Chapter 7.

6. K. Park, Limnol. Oceanog., 14, 179 (1969).

7. D. Dyrssen, Acta Chem. Scand., 19, 1265 (1965).

8. D. Dyrssen and L. G. Sillén, Tellus, 19, 110 (1967).

9. G. Gran, Analyst, 77, 661 (1952).

10. J. M. Edmond, Deep-Sea Res., 17, 737 (1970).

5

CHLORINITY

5.1 Introduction

Inasmuch as the ratios among the concentrations of the major ionic constituents remain essentially constant, it is possible to characterize the variation in water composition in terms of single parameter, typically the chlorinity or the salinity.

Salinity (S^o/oo) was originally defined as the weight in grams (*in vacuo*) of the solids that can be obtained from 1 kg of sea water (also measured *in vacuo*) when all of the carbonate has been converted to oxide, the bromine and iodine replaced by chlorine, all organic matter oxidized, and the residue dried at 480°C to constant weight.[1] In practice, the salinity is usually a secondary quantity and is defined in terms of chlorinity ($C\ell^o/oo$) and a standard relationship[1]

$$\text{Salinity } S^o/oo = 1.80655 C\ell^o/oo \tag{5-1}$$

Chlorinity (expressed in grams per kilogram of sea water sample) is defined as the mass in grams of pure silver necessary to precipitate the halogens in 0.3285233 kg of sea water.

Chlorosity (expressed in grams per liter of sea water) is the weight of chloride and bromide in a liter of sea water at 20°C, the bromide being replaced by an equivalent amount of chloride. The term is less commonly used.

The chemical and physical methods that have been used to determine salinity and chlorinity have been summarized elsewhere.[1-3] Two chemical methods have been selected here: the Mohr-Knudsen method and the potentiometric titration method.

5.2 Mohr-Knudsen Method

This consists in titrating the halide ions in sea water with standard silver nitrate solution, using a potassium chromate indicator. The pertinent chemical reactions are

$$Cl^- + Ag^+ \rightarrow \underline{AgCl}, \quad Br^- + Ag^+ \rightarrow \underline{AgBr} \qquad (5\text{-}2)$$

$$CrO_4^{2-} + 2Ag^+ \rightarrow Ag_2CrO_4 \qquad (5\text{-}3)$$

When a slight excess of silver ion is present, red silver chromate is formed, and the endpoint is indicated by a faint red color which persists in the solution for 30 sec. Additional details concerning procedures, precautions, and practices (especially for shipboard use) may be found elsewhere.[2-4]

The Mohr-Knudsen method is an empirical method, albeit one that is highly standardized. Considerable attention must be given to such details as the following. The titrations are carried out using a special Knudsen automatic pipet and buret; these must be carefully cleaned and calibrated. The temperature should be maintained as nearly constant as possible. The time required to analyze samples of approximately the same chlorinity should be nearly the same. The sample should be agitated during the titration by means of a magnetic stirrer. The rate must be fast enough to prevent formation of curds in the precipitate, but slow enough to prevent splashing.

Procedure

(a) Apparatus

Knudsen Buret Assembly. This is represented in Figure 5-1. The assembly consists of a black silver nitrate bottle (about 10 liters) (Note 1), Knudsen buret, magnetic stirrer, Knudsen pipet, and silver nitrate waste jar. The Knudsen buret is a delicate, precise, and expensive piece of volumetric apparatus. This buret differs from conventional burets inasmuch as it is calibrated in double milliliters, which permits larger amounts of a more dilute solution to be used, with an increase

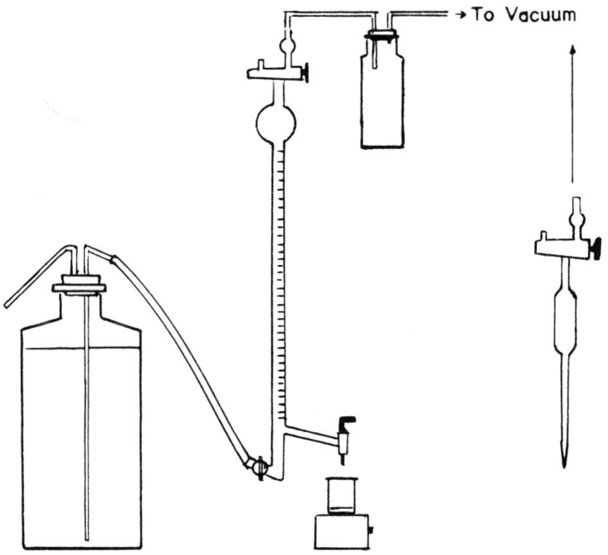

FIG. 5-1

Knudsen buret assembly.

in accuracy. A reading of 20.00 on the buret is equal to 40.00 mℓ of solution.

(b) Chemicals

Potassium Chromate Indicator. Dissolve 8 g of reagent-grade potassium chromate in 100 mℓ of distilled water. Store in a clean dropping bottle.

Silver Nitrate Solution. Dissolve 37.11 g of silver nitrate in a liter of distilled water (Note 2). If possible, prepare a 10-liter sample by weighing 371.1 g of silver nitrate in a clean 1000-mℓ beaker. Add 500 mℓ of distilled water, dissolve as much silver nitrate as possible, and transfer to the clean silver nitrate bottle. Repeat with additional 500-mℓ portions of distilled water, until all of the silver nitrate is dissolved, then dilute to 10 liters of solution. Stopper the silver nitrate bottle with a clean rubber stopper and shake the bottle vigorously for five minutes (Note 3).

Standard Sea Water (Eau de Mer Normale). This is the basis for all chlorinity titrations and is used to standardize the silver nitrate before and during a set of titrations. Because standard sea water is expensive, it is used with care, and a "sub-standard" sea water sample is prepared and compared with the standard sample. The points of the ampoule of standard sea water are cut with a glass knife. A clean, dry, 300-mℓ glass-stoppered bottle is washed with two small portions of the water and then filled from the ampoule. Drops of water which may adhere to the inner neck of the bottle should be removed with filter paper. Label the bottle indicating the chlorinity and the date. When a new ampoule must be opened, use a different glass-stoppered bottle.

Substandard Sea Water. Obtain about eight liters of sea water which has a chlorinity greater than 19.39°/oo. Filter the sample twice, using a No. 1 Whatman filter paper. Store the filtered water in a clean, well-leached black or amber bottle. After the chlorinity has been determined, calculate the volume of distilled water that must be added to convert the chlorinity to 19.39°/oo (Note 4). Add this volume of distilled water and shake vigorously for five minutes. Check the adjusted chlorinity against standard sea water.

(c) Sampling Water for Analysis

The sample and apparatus should be allowed to come to room temperature (during a six- or twelve-hour period). The pipet should be cleaned to allow for free drainage and must be calibrated before use. The pipet is mounted upright and the body of the pipet is not touched during any manipulation.

Rinse the pipet with the sample of the liquid to be titrated and repeat for each sample. This is done as follows. Place the dried tip of the pipet in the sample, and turn the threeway stopcock to the fill position. Fill the pipet, then turn the stopcock to allow the sample to drain into a waste jar. Repeat.

Collect a sample to be titrated and allow the sample to drain into a clean, dry, 100-mℓ beaker which contains a clean, plastic-covered magnet. Once the pipet has drained, allow the surface of the liquid to touch the tip of the pipet. Do not force any liquid from the pipet. Wipe the tip of the pipet with a scrap of filter paper or with a piece of lint-free absorbent tissue.

(d) <u>The Titration</u>

Perform the titration as follows. Add six drops of indicator solution to the sample in the beaker (Note 5). Place the beaker on the magnetic stirrer and beneath the delivery tip of the buret. Turn the rheostat of the stirrer to a setting low enough to preclude splashing. Open the delivery-stopcock and allow the solution to drain to the bottom of the bulb. Do not watch the buret from this point on; concentrate on the rate of stirring and delivery.

The color of the solution will pass through several shades of yellow and then become red. The first rough color change comes when the entire solution becomes a peach color, which will persist for 20 sec or less. Stop the addition of silver nitrate solution until the color becomes a pale yellow. Start adding silver nitrate dropwise; the solution in the beaker should become red where the silver nitrate hits. Then, add fractions of drops by opening the stopcock cautiously and closing it before a full drop can be formed. The droplet may be washed into the beaker with a stream of water from a wash bottle or the droplet may be transferred into the beaker with the tip of a clean stirring rod.

Repeat this technique until the endpoint is reached when the entire sample becomes a peach or dirty orange color. This color should persist for 30 sec during vigorous stirring.

Read and record this buret reading. Remove the stirring bar with a pickup rod and pour the contents of the beaker into the waste jar. Repeat the titration using another 15-mℓ volume of the sample. Buret readings should agree within 0.02.

(e) Calculations

The first step is the standardization of the silver nitrate solution against standard sea water. In effect, this consists in determining α, the difference between the known concentration (N) of the standard sea water and the unknown concentration (A) of the silver nitrate solution:

$$\alpha = (N) - (A) \tag{5-4}$$

For example,

(N)	Chlorinity of standard sea water	=	19.373
(A)	Buret reading from titration	=	19.510
	α	=	-0.137

Obtain the value of an average α from three titrations of standard sea water. Typical data are as follows:

(N)		(A)		α
19.373	-	19.510	=	-0.137
19.373	-	19.520	=	-0.147
19.373	-	19.525	=	-0.152

Alpha may be positive or negative, but the value must be in the range $-0.150^{o}/oo$ to $0.145^{o}/oo$. Otherwise, Knudsen's tables cannot be used and the silver nitrate solution must be adjusted (Note 6) by adding more distilled water ("strong" solution, positive α) or by adding more silver nitrate ("weak" solution, negative α).

The second step is to calculate the chlorinity from the titration data and after α has been determined. Use Knudsen's tables,[5] pp. 23-34, to determine the correction k which is algebraically added to the average buret reading to give the chlorinity. Find the page that has the observed value of α (top of page) and be sure that the sign agrees with that of the observed α. Find the average buret reading in the column below α; the sign and value of k will be found on the same line in the right-hand margin.

For example,

Alpha is -0.145

Average buret readings	20.110
Correction k (p. 23)	-0.16
$C\ell^{o}/oo$ is	19.95
S^{o}/oo is	36.04

Calculate the salinity (S^{o}/oo) from the formula given earlier; or better, use Knudsen's tables,[5] pp. 1-22.

5.3 <u>Potentiometric Titration Method</u>

Dyrssen and Jagner[6] have developed a potentiometric titration for determining chlorinity in sea water. The method is a practical one and computer calculation of results has been developed.

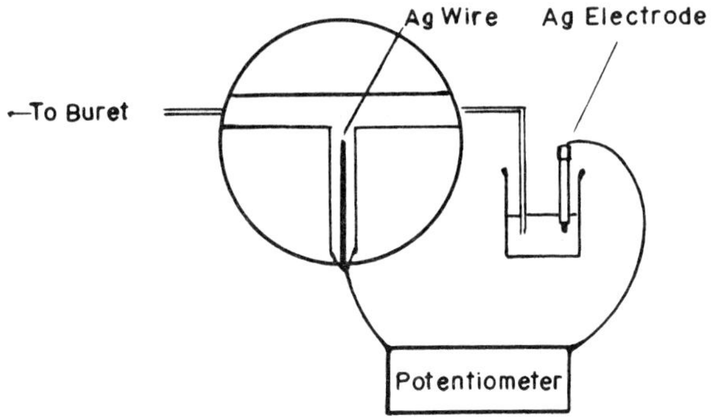

FIG. 5-2

Schematic drawing of potentiometric titration assembly, with magnified view (encircled) of standard solution of silver chloride.

Procedure[6]

(a) Apparatus

The apparatus is illustrated schematically in Figure 5-2. Note that one electrode is always in contact with the titrant, the other is in contact with the sample. Contact between the electrodes is maintained by having the titrant outlet below the surface of the sample. The titrant is delivered by a constant-rate buret or an automatic pipet, and the emf is measured by a suitable potentiometer (Note 7). The sample is titrated in a 100-mℓ beaker equipped with a plastic-coated stirring bar. The titrant is 0.2 M $AgNO_3$.

(b) Chemicals

Silver Nitrate (0.2 M). Dissolve 33.98 g of silver nitrate in distilled water and dilute to one liter (Note 2).

The solution is best prepared in 10-liter quantities. Store in a blackened bottle and shake well before use. Standardize the solution by using standard sea water.

(c) The Titration

Weigh or measure a sample of sea water using a 5-mℓ analytical-grade pipet. Dilute the sample with a known volume of water (25 mℓ) so that the salinity is about $5^{o}/oo$. Record the temperature. Turn on the magnetic stirrer. Add silver nitrate solution from the buret until the equivalence point is just reached as indicated by a sudden increase in the potential difference, usually at about 200 mV. Now, record volume and emf at 0.100-mℓ intervals. Continue until the difference in emf between two successive additions is less than 1 mV. Record the temperature, and use the average value in calculations.

Typical data are recorded in Table 5-1.

(d) Calculations

Determine the equivalence point by the graphical method. Plot the function $F = (v_o + v)10^{(E - E_1)/a}$ versus v (Note 8) where

v_o = initial volume

v = volume of standard silver nitrate solution added

E_1 = an arbitrarily chosen constant

a = RT log 10/F

R = 8.314×10^3 mV C deg^{-1} $mole^{-1}$

F = 96,520 C $mole^{-1}$

T = average absolute temperature during the titration

= °C + 273

log 10 = 2.303

After the equivalence point, F varies linearly with v. Draw

TABLE 5-1

Data for Sample Calculation of Chlorinity

v (mℓ)[a]	E (mV)	F x 10^{-2}[b]
00.0	458.0	—
13.98	340.0	—
14.18	205.0	0.06
14.21	181.0	0.926
14.24	170.0	1.422
14.28	157.5	2.320
14.33	147	3.49
14.38	140	4.60
14.48	131	6.56
14.58	122	9.31
14.64	118	10.9
14.70	115	12.26
14.78	111.5	14.08
14.84	109.3	15.70
14.90	107	16.90

[a]Milliliters of 0.1989 M $AgNO_3$ added.

[b]$F = (30.00 + v)10^{(E_1 - E)/a}$; a = 59.1, E_1 = 200 mV.

the best straight line through the linear portion and extrapolate to F = 0 to obtain the intercept on the x-axis $v = v_e$ (see Figure 5-3).

Calculate the chlorinity from the formula

$$Cl^{o}/oo = \frac{v_e \times N}{1000} \times \frac{107.868 \times 328.5233^{o}/oo}{w} \qquad (5\text{-}5)$$

where N = normality of the silver nitrate solution

w = weight of sample of sea water

Calculate the salinity as before.

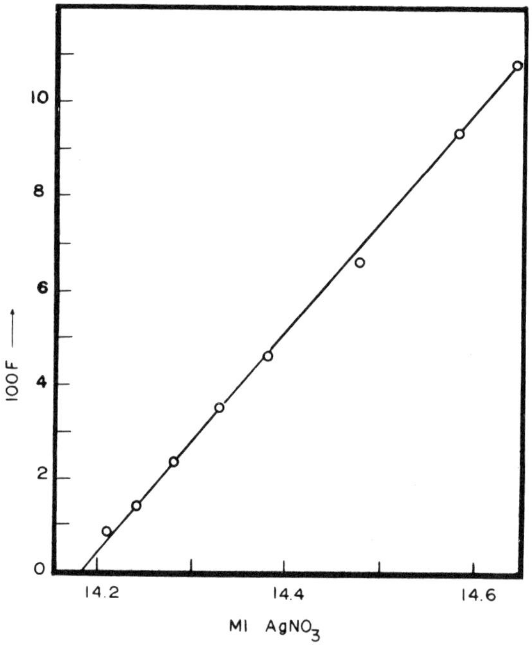

FIG. 5-3

Typical chlorinity titration plot using data in Table 5-1.

(e) Sample Calculation

The sample was drawn from 180 meters on 29 March 1967 during a cruise of the R/V Hernan Cortez. A 5.0 mℓ sample

(5.012 g) was diluted with 25 mℓ of distilled water and titrated with 0.1989 N silver nitrate. The data are recorded in Table 5-1.

For this titration, temperature = 24°C, a = 59.1, $F = (v_o + v)10^{(E_1 - E)/a}$, v_o = 30.0 mℓ, E_1 = 200 mV.

From the plot of F versus v (Figure 5-3), the value of v_e = 14.185 was obtained. Thus,

$$C\ell^o/oo = \frac{14.185 \times 107.8 \times 328.5233 \times 0.1989}{1000 \times 5.012}$$

$$= 20.015^o/oo$$

$$S^o/oo = 36.16^o/oo$$

NOTES

1. Black polyvinyl chloride tubing should be used for all connections.
2. Silver nitrate is a poisonous compound, both internally and externally. Anyone performing titrations should frequently wash their hands in salt water, then rinse with fresh water.
3. The bottle should be shaken vigorously for five minutes each day before the start of a series of titrations. At this point, the Knudsen buret should be filled with silver nitrate solution. When the buret has been standing for some time exposed to light, the solution should be drained and refilled twice. The buret should be free of air bubbles and should drain freely.

4. Let y equal the number of liters of distilled water that must be added to eight liters of sea water which has a chlorinity of $20.150^{o}/oo$. Then $y = (\frac{20.15}{19.39})8$ - eight liters.

5. Some workers prefer to use phenosafranin indicator. This is prepared as follows.[4] Add 0.25 g of sodium benzoate to 200 mℓ of distilled water in a 500-mℓ beaker, and bring to a vigorous boil. Meanwhile, prepare a suspension of 10 g of soluble starch in 20 mℓ of distilled water. Slowly, with stirring, add the cold starch suspension to the boiling sodium benzoate solution and continue to boil for 3-5 min. Filter the hot material through Pyrex glass wool. Finally, add 15 mℓ of distilled water to 0.8 g of phenosafranin concentrate, stir, and slowly, with stirring add the solution to the filtered, hot starch solution. Allow the mixture to cool to room temperature, then dilute to 250 mℓ.

   When phenosafranin indicator is used, add 1 mℓ to the sample. As silver nitrate solution is added, the color changes from red to pink to lavender. When the last color is observed, add the silver nitrate dropwise. The endpoint occurs when the solution is a solid blue color that persists for 20 sec. Should the lavender color return within that time, add a part of a drop of silver nitrate solution.

6. When α is negative, the weight (in grams) of silver nitrate to be added to each liter of solution is equal to $(2.01\alpha - 0.28)$ g. When α is positive, the milliliters of distilled water to be added to each liter of solution is equal to $(52.5\alpha - 0.4)$ mℓ.[4]

7. It should be possible to read the potentiometer to 0.6 mV (0.01 pH unit) for best results.[6]

8. If the electrodes are connected in such a way that the emf increases with increasing volume of titrant, use the function $F = (v_o - v)10^{(E - E_1)/a}$.

REFERENCES

1. J. Lyman, Limnol. Oceanog., 14, 928 (1969).

2. F. Hermann, K. Kalle, F. Koczy, W. Maniece, and P. Tchernia, J. Cons. int. Explor. Mer, 24, 429 (1959).

3. J. P. Riley, in Chemical Oceanography (J. P. Riley and G. Skirrow, eds.), Academic Press, New York, 1965, Chapter 21.

4. Instruction Manual for Obtaining Oceanographic Data, 3rd ed., U. S. Oceanographic Office, Washington, D. C., 1968.

5. M. Knudsen (ed.), Hydrographical Tables, Reported by Tutein & Koch, Copenhagen, 1959.

6. D. Dyrssen and D. Jagner, Anal. Chim. Acta, 35, 407-409 (1966).

6

SALINITY

6.1 Introduction

The method of salinity determination selected is often governed by the precision of the method, and the precision level that is appropriate is governed by the purpose and the location.

Consider the location first. A precision of $\pm 0.003^{o}/oo$, now standard in analysis of deep-sea waters, would represent wasted effort in analysis of estuary water. Even in an area of relatively gradual salinity changes, such as the middle bay of Chesapeake Bay, salinity is sufficiently a local and transient parameter as to preclude the need for good precision. For example, Mangelsdorf noted[1] that in the middle bay in summer, a precision of $\pm 0.003^{o}/oo$ would be wasted unless the position of sampling "were specified to within 40 feet, the time specified to within 90 seconds, and the depth specified to within a quarter of an inch!"

Next, consider the purpose. Salinity is used as a quantitative tool for tracing water masses and it is also used for qualitative studies, e.g., as an ecological parameter.

For quantitative studies, the appropriate precision varies with location: $\pm 0.1^{o}/oo$ or even $\pm 0.01^{o}/oo$ for estuaries to $\pm 0.003^{o}/oo$ or better for deep-sea oceanography. Conductivity is perhaps the best choice if the last two levels of precision are required. The limitations of conductivity and how they have been overcome has been discussed by Mangelsdorf[1] and Cox[2]. Several precision conductivity salinometers now in use have been described.[1-3]

The conductivity-chlorinity relationship in sea water is

$$\text{C}\ell\,^{o}/oo = -\ 0.04980 + 15.66367R_{15} + 7.08993R_{15}{}^{2} - 5.91110R_{15}{}^{3} + 3.31363R_{15}{}^{4} - 0.73240R_{15}{}^{5} \quad (6\text{-}1)$$

Here, R_{15} is the ratio of the electrical conductivity of a given sample to that of one S = 35°/oo with both samples at 1 atm and 15°C.

Other precise direct determinations of salinity have been made by direct gravimetric determination,[5] by refractivity (Note 1),[6] and by a freeze-drying technique.[7]

At another, lower, level of precision, salinity is used as an ecological parameter that helps define limits of distribution of species. Salinities measured with a relative precision of 5% are adequate for this purpose. This means ±1°/oo for S = 20°/oo and ±0.1^{o}/oo for S = 2°/oo. At this precision level, salinity can be determined readily by means of specific gravity using a precision hydrometer, by means of refractive index, and by means of a chloride titration.

Two rapid and fairly precise methods are included here: direct determination of salinity by refractivity and by Harvey's method.[8]

6.2 Refractive Index Method

The refractive index of a medium, such as sea water, is defined as the ratio of the speed of light in a vacuum to the speed in the medium. The refractive index of sea water is a useful property to measure salinity because the refractive index is a function of the salinity,[6] because the effect of temperature variation on refractive index is small, and because the effect of temperature on refractive index does not vary significantly with salinity.[7] The refractive index method has two requirements: a precise refractometer and comparison of an unknown sample of sea water with a sample of known salinity at the same temperature. The first requirement arises because an increase in salinity of $1^{o}/oo$ is accompanied by an increase in the refractive index by about 0.0002. The second requirement arises from a need to minimize the effect of temperature on refractive index and increase the precision.

Salinity can be determined to about $0.05^{o}/oo$ with a refractometer precise to 0.00001. In order to obtain higher precision, a comparison instrument would need to be used, e.g., a differential prismatic refractometer or an interferometer. Such instruments would have a working accuracy of $0.01^{o}/oo$. For practical work, a hand-held refractometer can be used with an accuracy of $0.5\text{-}0.6^{o}/oo$.

Procedure

(a) Apparatus

Refractometer. A hand-held refractometer, temperature

compensated for direct reading of serum protein and refractive index, is useful for laboratory and field work (Note 2).

(b) Reagents

Salinity Samples. Distilled water and a sea water of known salinity are needed. These should be stored in tightly sealed bottles equipped with medicine droppers (Note 3).

(c) Determination of Salinity

The water sample should be filtered through Whatman No. 1 paper, if much suspended matter is present.

Determine and record the refractive index of the unknown sample, distilled water, and the standard salinity sample at the same temperature, following the instructions that accompany the refractometer.

For precise work, a calibration plot (refractive index as a function of salinity) could be prepared at the working temperature.

The slope of the calibration line is given by the relationship

$$\text{slope} = \frac{\text{reading for standard salinity} - \text{reading for distilled water}}{\text{salinity of standard sample}} \quad (6\text{-}2)$$

The salinity of the unknown sample may be found from the relationship

$$\text{salinity of unknown} = \frac{\text{reading for unknown sample} - \text{reading for distilled water}}{\text{slope of calibration line readings}} \quad (6\text{-}3)$$

For example, the following readings were obtained:

distilled water:	1.3330
standard salinity sample (S, 34.998°/oo):	1.3395
unknown sample:	1.3387

For these readings, the slope is equal to

$$\frac{1.3395 - 1.3330}{34.998} = 1.86 \times 10^{-4}$$

The calculated salinity is equal to

$$\frac{1.3387 - 1.3330}{1.86 \times 10^{-4}} = \frac{57 \times 10^{-4}}{1.86 \times 10^{-4}} = 30.7^{\circ}/oo$$

6.3 The Harvey Method[7]

A simple titrimetric method for determining salinity has been given by Harvey.[7] The accuracy is less than that obtained by the procedures given in Chapter 5, but the accuracy is sufficiently good for determining the salinity of tidal pools and culture media.

Procedure

(a) Reagents

Silver Nitrate Solution. Dissolve 27.25 g of silver nitrate in distilled water and dilute to one liter. Use the precautions given in Chapter 5.

Potassium Chromate Indicator. Dissolve 8 g of reagent grade K_2CrO_4 in 100 mℓ of distilled water. Store in a clean dropping bottle.

(b) Analysis

Use a 50-mℓ buret that can be read to the nearest 0.02 mℓ and titrate a 10-mℓ sample of water to which has been added four drops of chromate indicator. Follow the general procedure of the Mohr-Knudsen procedure (Chapter 5). Record the volume of standard silver nitrate solution needed to reach the end-point.

(c) Calculations

The salinity (in parts per thousand) is almost numerically equal to the volume (in milliliters) of silver nitrate solution needed to titrate a 10-mℓ sample.

A more accurate value can be obtained by adding a small correction to the observed salinity of the buret reading (Note 4).

For example, a 10-mℓ sample required 36.05 mℓ of standard silver nitrate solution. The observed salinity is $36.05^{o}/oo$. The corrected salinity is 36.05 + (-0.03), or $36.02^{o}/oo$. If the observed salinity is $30.05^{o}/oo$, the corrected salinity is (30.05 + 0.11), or $30.16^{o}/oo$.

It is better practice to determine the corrections for the buret and pipet being used. This is done by titrating samples which have known values of salinity in the range of interest and comparing the buret reading with the known salinity.

NOTES

1. Technically, refractivity was found to be a linear function of chlorinity.[6]

2. One such instrument is the AO-TS meter, Model 10401, which is manufactured by the American Optical Company,

Mechanic Street, Southbridge, Mass.

3. A range of salinities can be prepared by careful dilution (Chapter 5).

4. For the following observed salinities, the corrections[8] to be applied are given in parenthesis: 40 (-0.15), 38 (-0.08), 36 (-0.03), 34 (0.03), 32 (0.07), 30 (0.11), 28 (0.15), 26 (0.17), 24 (0.20), 22 (0.22), 20-16 (0.24), 14 (0.12), 12 (0.19), 10 (0.16), 8 (0.15).

REFERENCES

1. P. C. Mangelsdorf, Jr., in Estuaries (G. H. Lauff, ed.), AAAS, Washington, D. C., 1967, p. 71.

2. R. A. Cox, in Chemical Oceanography (J. P. Riley and G. Skirrow, eds.), Vol. 1, Academic Press, New York, 1965, pp. 103-107.

3. Instruction Manual for Obtaining Oceanographic Data, 3rd ed., U. S. Oceanographic Office, Washington, D. C., 1968, p. I-1.

4. J. Lyman, Limnol. Oceanog., 14, 928 (1969).

5. A. W. Morris and J. P. Riley, Deep-Sea Res., 11, 899 (1964).

6. C. L. Utterback, T. G. Thompson, and B. D. Thomas, J. con. int. Explor. Mer, 9, 35 (1934).

7. J. Lyman, R. F. Barquist, and A. V. Wolf, J. Mar. Res., 17, 335 (1958).

8. H. W. Harvey, The Chemistry and Fertility of Sea Waters, 2nd ed., Cambridge University Press, London, 1957, p. 127.

7

DISSOLVED OXYGEN CONTENT

7.1 Introduction

The concentration of dissolved oxygen in sea water varies markedly, from zero to in excess of ten milliliters per liter of sea water. The former value would be found in stagnant waters, the latter in a region of supersaturation, e.g., at the surface and in the presence of great photosynthetic activity. The determination of dissolved oxygen content is important for a variety of reasons, the most prominent of which is that a knowledge of oxygen content is invaluable in understanding biochemical processes which occur in the ocean.

It is true that physical methods have superseded chemical methods of analysis in many instances. This is less true in the determination of dissolved oxygen; though polarographic methods have been developed,[1] calibration with solutions of known oxygen content is recommended.

The classical chemical method, the Winkler titrations, may be used to determine dissolved oxygen concentration in natural waters that are relatively free of strong oxidizing or reducing agents (nitrites, ferrous salts, organic materials). This condition is satisfied in most areas of the open sea, but with polluted waters, modifications must be applied (Note 1). The Winkler method consists in allowing manganous hydroxide to react with the dissolved oxygen and be oxidized to an insoluble brown manganese (IV) compound; when the resulting solution is acidified, in the presence of excess potassium iodide, iodine is formed quantitatively and may be titrated with a standard solution of sodium thiosulfate.

The pertinent reactions may be represented by the equations

$$Mn^{2+} + 2OH^- \rightarrow Mn(OH)_2 \text{ (white)} \tag{7-1}$$

$$2Mn(OH)_2 + O_2 \rightarrow 2MnO(OH)_2 \text{ (brown)} \tag{7-2}$$

$$MnO(OH)_2 + 4H^+ + 3I^- \rightarrow Mn^{2+} + I_3^- + 3H_2O \tag{7-3}$$

$$I_3^- + 2S_2O_3^{2-} \rightarrow 3I^- + S_4O_6^{2-} \tag{7-4}$$

The problems involved in the use of the Winkler method include the following.[1,2]

Sampling. The analyses must be started within an hour of collection of the sample. Corrosion of metal sampling devices can absorb up to 0.04 mℓ O_2/liter per hour. This is obviated by use of plastic or plastic-coated bottles. Even so, oxygen loss by temperature increase or pressure loss or by biochemical process can occur.

Volatilization of Iodine. This can occur if the samples are stored in the acidic state. Grasshoff[3] recommends use of 70 g of potassium iodide/100 mℓ, instead of the lower amounts usually recommended; this increases the tendency to form triiodide ion and reduces the danger of volatilization of iodine.

Standardization of Potassium Thiosulfate. Typically, potassium biniodate or potassium dichromate is used to standardize the thiosulfate solutions. It appears that there is a systematic error in the use of potassium dichromate which can lead to low values for the thiosulfate normality: the reaction between dichromate and iodide is rapid only in the presence of large concentrations of acid and iodide, and under these condition, photochemical oxidation of iodide tends to occur.

End-Point Errors. Knowles and Lowden[4] found that the use of starch indicator can lead to results that are on the low side, but they devised an amperometric titration that greatly reduces the end-point error.

7.2 The Winkler Determination

The following procedure describes a conventional technique; entire-bottle technique is described in Section 25.4.

Procedure

(a) Reagents

Manganous Sulfate Solution. Dissolve 450 g of $MnSO_4 \cdot 4H_2O$ in water and dilute to one liter. The manganous sulfate must be free of ferric compounds; otherwise, it should be recrystallized.

Alkaline Iodide Solution. Dissolve 700 g of potassium hydroxide or 500 g of sodium hydroxide in 500 mℓ of distilled water; and 150 g of potassium iodide in 200 mℓ of distilled water. Add the potassium iodide solution to the potassium hydroxide solution and dilute to one liter. Store this solution in an amber glass bottle with a rubber stopper. Keep the solution out of contact with sunlight.

Starch Solution.[5] Prepare a slurry of 5 g of "soluble" starch in 30 mℓ of formamide at room temperature. Meanwhile, heat 65 mℓ of formamide (using a hood) to 100-110°C. Pour the slurry with stirring into the hot formamide solution. The starch should dissolve within a minute and the solution is ready for use once it is cool. Holler[5] reported that the solution is stable for at least eight months (Note 2).

Concentrated Sulfuric Acid (18 M, 36 N). Use 96% acid (sp. gr. 1.84).

Sodium Thiosulfate Solution. Dissolve 3.5 g of $Na_2S_2O_3 \cdot 5H_2O$ in one liter of distilled water that has been freshly boiled for about ten minutes to dispel carbon dioxide. Store in an amber bottle, and add five pellets of sodium hydroxide per liter of solution. Allow the solution to age for several days before standardizing.

Potassium Biniodate Solution Accurately weigh a sample, w grams of $KH(IO_3)_2$ (about 0.325 g), dissolve in distilled water, and dilute to one liter. The normality N_1 of the sample is w/32.5.

(b) The Blank Run

Before determining the normality of the thiosulfate solution, it is necessary to correct for the presence of trace impurities in the reagents. This must be done each time a new set of reagents is prepared, or weekly, depending upon the amount of use. The blank run is conducted by preparing and testing a blank solution.

Prepare a blank solution in the following way. Carefully, measure 97 mℓ of freshly boiled distilled water (using a 100-mℓ graduated cylinder), and pour the sample into a 250-mℓ glass-stoppered Erlenmeyer flask. Add successively with mixing: conc H_2SO_4 (1 mℓ), alkaline iodide reagent (1 mℓ), manganous sulfate (1 mℓ).

The reagent solutions are free of oxidizing agents if a blue coloration does not develop when two drops of starch solution are added. Otherwise, determine the volume of thiosulfate needed to dispel the color. This volume V_b should be less than 0.10 mℓ. If this is not the case, the impure reagent must be detected and eliminated.

(c) Standardization of Sodium Thiosulfate

First, prepare a blank solution as before. Then, add 25 mℓ of potassium biniodate solution with a pipet. Swirl gently, restopper the flask, and allow the sample to stand in the dark for about ten minutes.

Titrate the iodine with thiosulfate solution, using a 25- or 50-mℓ buret. When the solution is a pale straw color, add two drops of starch solution and titrate to the disappearance of the blue color. Record the volume required to the nearest 0.02 mℓ.

Repeat the standardization at least twice and average the best results.

The normality of the thiosulfate solution is N:

$$N = \frac{V_1 N_1}{V_2 - V_b} \tag{7-5}$$

where

V_1 = volume of standard biniodate solution; this is always 25.0 mℓ

N_1 = normality of the standard potassium biiodate solution (= w/32.5)

V_2 = milliliters of sodium thiosulfate used

V_b = average milliliters correction, obtained from the blank run

(d) Treatment of Seawater Sample

Samples for oxygen analyses must be withdrawn before any other samples are removed from the water sampling bottles. Preserve samples to be analyzed in 250-mℓ amber reagent bottles with glass tops. The following steps are involved in withdrawing a sample.

Rinse the bottle. Add a small amount of water to the bottle, shake vigorously, and pour the rinse water over the glass stopper. Repeat. This should displace the air from the 20-30-cm length of rubber tubing which is attached to the drain petcock. Open the petcock slowly to prevent bubbles of air from entering the sampler.

Fill the sample bottles. Place the end of the rubber tubing in the bottom of the amber bottle. Fill the bottle slowly so as to avoid air bubbles and allow the sample bottle to overflow. Carefully withdraw the delivery tube, then turn off the petcock. Quickly, but carefully, replace the ground glass stopper in such a way as to prevent bubbles of air from being entrapped in the bottle.

Treat the samples as quickly as possible with manganous sulfate and alkaline iodide solutions. Use a pipet to introduce 1 mℓ of manganous sulfate well below the surface of the water. Add 1 mℓ of alkaline iodide in the same manner. Stopper the bottle quickly, but avoid entrapping air bubbles; ignore any overflow. Shake the bottle vigorously for a minute to mix the reagents, and after the precipitate has settled, shake again.

(e) Analysis of the Sample

The treated samples should be analyzed after the precipitate has had a chance to react with the dissolved oxygen and to settle to the bottom of the bottle. This requires at least 30 minutes, but the samples should be analyzed within six hours. Using a pipet, introduce 1 mℓ of concentrated sulfuric acid well underneath the surface of the sample. Replace the stopper and shake until all of the precipitate is dissolved and the iodine is liberated.

Remove a 100-mℓ sample (preferably with a pipet rinsed with sample) and transfer to a clean 250-mℓ Erlenmeyer flask which is equipped with a clean, plastic-coated magnetic stirring bar.

Titrate the solution with standardized sodium thiosulfate until the solution is a very pale yellow. Add three drops of starch indicator and continue the titration until the color disappears. Record the volume V of standard thiosulfate solution added. Drain the flask and repeat the titration using a second 100-mℓ sample. Two titrations should agree within ±0.03 mℓ. If necessary, run a third titration with a 50-mℓ sample.

(f) Calculations

The volume of the treated sample is 100 mℓ, but the volume of sea water actually used is

$$100 \times \frac{(250 - 2)}{250} = 99.2 \text{ m}\ell$$

For this sample, the oxygen content, expressed in units of milliliters of dissolved O_2 per liter of sea water, is

$$56.45N\,\overline{V} \text{ m}\ell\ O_2 \text{ at STP/liter (Note 3)}$$

Here, N is the normality of standardized sodium thiosulfate solution, and $\overline{V}$ is the average volume of standardized thiosulfate used to titrate a 100-mℓ sample.

NOTES

1. The details are given in Chapter 25.
2. Aqueous starch solution: Grind 0.5 g of soluble starch with small amounts of distilled water until a thoroughly fluid paste is obtained. Pour the paste into 200 mℓ of distilled water which is just at the boiling point. Boil for one minute. Daily preparation is recommended for aqueous starch.
3. This can be verified knowing that the milligram equivalent (meq) of thiosulfate is NV; 1 meq thiosulfate = 8 mg O_2;

and 1 mg O_2 at STP (0°C, 760 mm Hg) = 0.700 mℓ of O_2.

REFERENCES

1. J. P. Riley, in Chemical Oceanography (J. P. Riley and G. Skirrow, eds.), Academic Press, New York, 1965, Chapter 21.

2. H. Barnes, Apparatus and Methods of Oceanography Part 1: Chemical, George Allen & Unwin Ltd., London, 1959, Chapter 14.

3. K. Grasshoff, Kiel Meeresforsch, 18, 42-50 (1962); Chem. Abstr., 57, 13552 (1962).

4. G. Knowles and G. F. Lowden, Analyst, 78, 159-64 (1953).

5. A. C. Holler, Anal. Chem., 27, 866 (1955).

Section II

ANALYSIS OF SELECTED MICRONUTRIENTS AND RELATED COMPOUNDS

8

SPECTROCHEMICAL ANALYSIS OF MICRONUTRIENTS

8.1 Introduction

The growth of marine plants in the open sea is limited primarily by two environmental factors: light and nutrients. Obviously, the variation of these factors is of considerable interest, as are those hydrographic features that affect the variation (see Volume 2, Chapter 9). Several chapters that follow are devoted to the methods of analysis of trace constituents of sea water, including various forms of nitrogen and phosphorus and metals such as iron, copper, manganese, and molybdenum. These constituents are micronutrients and the analyses depend upon spectroscopy.

Three general analytical approaches can be used: optical, electrical, and other methods such as mass spectrometry, neutron activation, and the use of catalytic reactions. Optical methods include atomic absorption spectroscopy (Chapter 22), flame photometry (Chapter 23), and colorimetry (Note 1).

8.2 Colorimetry[1,2]

Spectrochemical methods typically depend upon the use of a chemical reaction to produce a colored species that can be analyzed by colorimetry (Note 1). The "success" of a procedure can be expressed in terms of the sensitivity of the detection limit. The "sensitivity" is that concentration that produces an absorption of 1% on the spectrometer; the "detection limit" is that concentration that produces an absorption equivalent to twice the background fluctuation (noise level).

Colorimetry has the advantages of being relatively simple, inexpensive, and adaptable to analyses of a variety of constituents. It is often extremely sensitive. For example, the estimated detection limits (as parts per billion) for iron using various optical methods are: 1100 (flame emission), 50 (atomic absorption spectroscopy), 15 (plasma emission spectroscopy, Note 2), and 2 (colorimetry).

There are, of course, disadvantages associated with colorimetry. A greater possibility of contamination exists than with a direct method (such as atomic absorption spectroscopy) because reagents (with trace impurities) are added, or because separation or concentration steps are involved. Also, as with many procedures, the sensitivity may vary markedly from one laboratory to another, as a consequence of differences in the analysts' skills and instrument qualities.

Optical methods are based upon detecting the interaction of light, or, properly, radiant energy with matter. The photon concept and the quantization of energy are two pertinent concepts.

The Photon Concept. Radiant energy consists of discrete units of energy called photons. The energy content (E) of each photon is a function of the frequency:

$$E = h\nu \tag{8-1}$$

where ν is the frequency (Note 3) and h is Planck's constant (a fundamental constant equal to 6.62×10^{-27} erg-seconds).

Quantization of Energy. Radiant energy can be absorbed and/or emitted by matter only in certain units or quanta. This is because matter (as molecules, atoms, or ions) can only have certain energy levels, and the transition from one energy level to another requires the absorption or emission of energy. A spectrum results from the absorption (or emission) of radiant energy at discrete frequencies, Eqn. (8-1) or at discrete wave lengths, λ,

$$\lambda = c/\nu \tag{8-2}$$

where c is the velocity of light (3×10^{10} cm/sec).

Several units have been used to express wavelength. These include angstrom units (Å) microns (μ), millimicrons ($m\mu$), and nanometers (nm):

$$1\ \text{Å} = 10^{-8}\ \text{cm} = 10^{-10}\ \text{m} \tag{8-3}$$

$$1\ \mu = 10^{-4}\ \text{cm} = 10^{-6}\ \text{m} \tag{8-4}$$

$$1\ m\mu = 10\ \text{Å} = 10^{-3}\ \mu = 10^{-7}\ \text{cm} = 10^{-9}\ \text{m} \tag{8-5}$$

$$1\ \text{nm} = 1\ m\mu = 10^{-9}\ \text{m} \tag{8-6}$$

Of these units, millimicrons have been used commonly in colorimetry, though nanometers are the preferred unit.

Table 8-1 shows diagrammatically some regions of interest to the marine chemist. The energies given are approximate and are expressed in calories per mole, i.e., the calories of energy absorbed (or emitted) when 6×10^{23} molecules undergo a given transition. The application of the two concepts can be

TABLE 8-1

Regions of the Electromagnetic Spectrum

Spectral region	Wavelength limits, Å	E, cal/mole	Type of transition
Far-infrared	2×10^5-10^7	1,500-30	Molecular rotation
Near-infrared	8000-2×10^5	37,500-1,500	Molecular vibrations
Visible	4000-8000	75,000-37,5000	Transitions of valence electrons; band spectra for molecules; line spectra for molecules
Ultraviolet	2000-4000	150,000-75,000	As for visible region
Vacuum-ultraviolet	20-2000	1.5×10^7-150,000	K and L electron transitions
X-ray	0.1-20	3×10^9-1.5×10^7	Nuclear transitions

seen best by considering the typical situation in a colorimetric experiment. One substance is dissolved in a colorless solvent and placed in a spectrophotometer cell which has transparent, plane, parallel faces. A beam of monochromatic radiation is passed through the solution along a distance b (path length). The radiant power of the beam (Note 4) will decrease as more increments of molecules absorb equal fractions of radiant energy (i.e., as b increases and as the concentration of molecules per unit volume increases).

The absorption of radiant energy by this substance (or matter in general) is described by Beer's Law (Note 5). The simplest statement of this law is

$$A = abc = \log_{10}(P_0/P) \qquad (8\text{-}7)$$

where A is the "absorbance" (sometimes called the optical density); b is the path length, usually expressed in centimeters; c is the concentration, usually expressed in molarity; and a is the absorptivity; P_0 and P are the radiant powers of the incident and emerging light, respectively (Note 4). (The symbols used with Beer's Law are given in Table 8-2).

According to Beer's Law, the absorptivity a is a constant which is independent of concentration. Nothing is said about the effects of wavelength, solvent, or temperature. In practice, Beer's Law is verified for a given concentration range at a specified wavelength, for a given solvent at room temperature and using a narrow-wavelength beam of light (as in a spectrophotometer).

TABLE 8-2

Symbols and Units Associated with Beer's Law

Symbol	Accepted Name	Definition
A	Absorbance	abc $\log_{10}(P_0/P)$
T	Transmittance	P/P_0
a	Absorptivity	A/bc
ε	Molar absorptivity (Molar extinction coefficient)	A/b $\underline{M}$
b	Path length	—

Beer's Law is said to be valid when a plot of absorbance A as a function of concentration is linear. Conformity is not necessarily a requirement for analysis provided the deviation is known from a calibration plot obtained under actual conditions.

Deviations from Beer's Law need to be considered, however, because it may be possible to design a better procedure. Some deviations are chemical in origin, e.g., the change in color of dichromate ion upon dilution which can be understood in terms of the equilibria (Note 6)

$$\underset{\text{(orange)}}{Cr_2O_7^{2-}} + H_2O \rightleftarrows 2H^+ + \underset{\text{(yellow)}}{2CrO_4^{2-}} \qquad (8\text{-}8)$$

Other deviations are physical in origin, e.g., the radiation may be too heterochromic. The effect is significant in some filter photometers, but not in a good spectrophotometer with a narrow wavelength light source.

8.3 Some Spectrochemical Techniques

Though available spectrophotometers differ widely, the procedures in use have certain common features. These are given in the next section. Cell-to-cell blanks are run to determine flaws in the cells. Direct absorption analysis is a standard colorimetric procedure, though the details will differ from instrument to instrument. Finally, the sensitivity may be improved by the relative photometry techniques that are given. These, in effect, expand the per cent transmittance scale.

Cell-to-Cell Blank. Fill both sample and reference cells with distilled water and measure the absorbance of the sample relative to the reference cell. The value should be 0.000, but there may be slight defects or the cells may need recleaning. If the recleaning does not change the cell-to-cell blank, a correction must be made (or another pair of better matched cells purchased).

Direct Absorption Analysis. Typically, the following steps are involved for all spectrophotometers:

1. Place the test solution in the sample cell or cuvette. Be certain the cells are made of appropriate material for the wavelength range to be used.

2. Select the appropriate photocell and lamp for the desired wavelength.

3. After a few minutes warm-up time, adjust the zero control (shutter closed, photocell dark) to "0" on the per cent

transmittance scale.

4. Turn the wavelength dial to the wavelength desired.

5. Place the reference cell in the measuring position and switch the shutter to the open position.

6. Adjust the instrument so the per cent transmittance dial reads 100% (zero absorbance).

7. Place the sample cell in the measuring position; determine and record the absorbance.

Relative Photometry.[3] When Beer's Law is followed, the differential or relative photometry procedure can be used (as in the analysis of manganese, Chapter 18) with a considerable enhancement of precision.

Use three solutions: (a) a "blank" which contains no test constituent but is otherwise treated the same as the sample; (b) a standard solution, which contains a known amount or a known increment of test constituent (c_s); and (c) the unknown solution, which contains an unknown amount of test constituent (c_x). The concentration in the unknown and standard solutions should be comparable.

Determine the absorbance of the unknown solution A_x relative to the blank.

It will be seen from the following that the relative absorbance $\log(P_x/P_s)$ is proportional to the difference in concentration between the unknown and standard solutions:

$$A_s = abc_s = \log(P_0/P_s) \qquad (8\text{-}9)$$

$$A_x = abc_x = \log(P_0/P_x) \qquad (8\text{-}10)$$

Subtracting (8-9) and (8-10),

$$\log_{10}(P_x/P_s) = ab(c_s - c_x) = A_s - A_x \tag{8-11}$$

There are three possibilities:

1. If a and b are known (Chapter 16), c_x can be calculated directly [Eqn. (8-11)].

2. If c_s is known, Eqns. (8-9) and (-10) are rearranged and solved for ab, and the unknown is calculated using

$$A_s/c_s = ab;\ A_x/c_x = ab$$

or

$$A_s/c_s = A_x/x_x;\ c_x = (A_x/A_s)\ c_s \tag{8-12}$$

3. If an increment of test constituent was added (Chapter 18): A slight modification is needed. Let c_i be the concentration of the increment of test constituent. Then the concentration of the standard solution is

$$c_s = c_i + c_x \tag{8-13}$$

This expression can be combined with Eqn. (8-12) and rearranged to give the following expression which is used in Chapter 18:

$$c_s = \frac{A_x}{(A_s - A_x)c_i} \tag{8-14}$$

Double Standard Method. This procedure was devised by Reilley and Crawford[4] and is an extension of the preceding method. Two standards are used, one slightly more concentrated than the other. This is potentially the most precise method of all colorimetric procedures. The method can be used when the test solution can be diluted.

8.4 Preparation of Deionized Water

Deionized water, suitable for nutrient analyses, can be

obtained from a mixed-bed ion exchange column (Note 7). Use Dowex 50W-x8 (50-100 mesh) cation-exchange resin (H^+ form) and Dowex 1-x4 (50-100 mesh) anion-exchange resin ($C\ell^-$ form), or equivalent.

First, clean about 25 mℓ of each resin by alternately washing with four column volumes of 4 M ammonium hydroxide (Note 7), followed by distilled water until effluent neutral to pH paper, then four column volumes of 5 M hydrochloric acid and distilled water (until neutral to pH paper). Repeat the cycle, then store each resin under distilled water.

Next, convert the cleaned anion resin to the chloride ion form. Place 10 mℓ in an ion-exchange column and pass about 500 mℓ of 3 M NaOH through the column at a flow rate of 2 mℓ/min. Continue eluting until a sample of the eluent shows an absence of chloride (acidify the sample with 1 M HNO_3, test with 0.1 M $AgNO_3$). Wash the resin water with 10-mℓ portions of distilled water until neutral (to pH paper).

Once the mixture is homogeneous, transfer it as a slurry to the anion-exchange column.

Prepare deionized water by passing distilled water through the mixed-bed ion-exchange column. Use a flow rate of 1 mℓ/min or less. Discard the initial 50-100 mℓ. Over 20 liters of deionized water can be collected without loss of water quality.

NOTES

1. Colorimetry properly applies in terms of the visible spectrum.[1]
2. Plasma emission spectroscopy uses an electrically generated plasma in nitrogen or other inert gas as the excitation source instead of the simple acetylene-oxygen flame photometry (Chapter 23). This is a new technique, still in the development stage, which offers promise of increased sensitivity in analyses of trace constituents.[5]
3. Frequency is the number of oscillations per second, in this instance, by the photon, which has some properties of a wave.
4. This is commonly called intensity I. According to Ewing[2] radiant power P is probably the more accurate term for the beam of radiant energy because of the units (energy per unit of time is power).
5. This is sometimes called the Beer-Lambert or the Bouguer-Beer Law. It is assumed that a correction for surface reflections and other losses is made by running a blank.
6. No deviation is found at 445 mμ because chromate and dichromate have the same molar absorptivity at this wavelength.
7. This procedure was described by Adamo.[6] Solutions: 4 M NH_3 (dilute 265 mℓ 28% ammonia to one liter); 5 M HCℓ (dilute 415 mℓ of 37% acid to one liter); 3 M NaOH (120 g sodium hydroxide diluted to one liter); nitric acid (dilute

6.3 mℓ of 69% acid to 100 mℓ); 0.1 M $AgNO_3$ (1.7 g diluted to 100 mℓ).

REFERENCES

1. D. N. Hume, Adv. Chem. Ser., 67, 30 (1967).

2. G. W. Ewing, Instrumental Methods of Chemical Analysis, 2nd ed., McGraw-Hill, New York, 1960.

3. C. F. Hiskey, Anal. Chem., 21, 1440 (1949).

4. C. N. Reilley and C. M. Crawford, Anal. Chem., 27, 716 (1955).

5. C. D. West and D. N. Hume, Anal. Chem., 36, 412 (1964).

6. F. S. Adamo, M. S. Thesis, Univ. of South Florida, 1968.

9

SILICON

9.1 Introduction

In natural waters, silicon is found in solution (as silicate and silicic acid) and in suspension (as silica, as clay minerals, and combined in organisms). The methods for determination of the soluble forms depend upon the reaction of orthosilicic acid (H_4SiO_4) with molybdate to form the 1:12 silicomolybdic acid. The absorbance of this intensely yellow colored product or the reduction product (molybdenum blue) is measured at selected wavelengths. According to Grasshoff and Hahn,[1] the formation of the silicomolybdic acid is reversible:

$$H_4SiO_4 + 2H_3(H_3Mo_6O_{21}) \rightleftarrows H_4(SiMo_{12}O_{40}) + 6H_2O \qquad (9\text{-}1)$$

The silicomolybdic acid exists as two modifications, the α and β species; the latter species is unstable and is converted into the α species.[2] The limitations of the determination are described elsewhere (Volume 2, Chapter 5).

9.2 Determination of Dissolved Silicon

The method used here was developed by Grasshoff[3] and involves two procedures: the "yellow" procedure, or formation of α-silicomolybdic acid (for waters containing large amounts of silica, i.e., in excess of 50μg-atom Si/liter), and the "blue" procedure, or reduction of the α species (for waters containing low amounts of silica). Grasshoff reported[3] the standard deviation as 0.12 μg-at Si/liter at the 10.8 μg-at level ("blue" procedure). At the 60 μg-at level, the standard deviations were 0.48 and 0.25 μg-at Si/liter for the "yellow" and "blue" procedures, respectively.

Procedure

(a) Reagents

Silica-Free Water. Distill deionized water in a quartz apparatus which has been used for some time (Note 1).

Standard Silicate Solution. Two silicate standards are available, sodium hexafluorosilicate, Na_2SiF_6, and pure quartz (suitably treated). The latter standard is less convenient to use and standard solutions are stable for about two months. The first standard is probably a less accurate one, but this is compensated by the convenience and the stability of standard solutions prepared from Na_2SiF_6.

Na_2SiF_6 Standard. Weigh out 1.920 g of dried (105°C) powder in a plastic beaker and dissolve in about 100 mℓ of silica-free water (Note 2). Transfer the contents of the beaker quantitatively to a one-liter volumetric flask (Note 3) and dilute to the mark. As soon as possible, transfer the standard solution to a plastic bottle; the solution rapidly picks up silica from glass. The concentration of this solution is 10 μg-at Si/mℓ.

Quartz Standard. Place pure quartz powder in a platinum crucible and ignite at 1000°C to constant weight. Weigh 600.6 mg of the dry material into a platinum crucible, mix well with about 5 g of anhydrous sodium carbonate, ignite the mixture to a clear melt at about 1100°C, cool to room temperature, and dissolve in silica-free water. Carefully transfer the solution to a volumetric flask and dilute to one liter with silica-free water. The solution contains 10 μg-at Si/mℓ, and should be stable for two months (Note 3).

Molybdate Reagent. Dissolve 120 g of sodium molybdate dihydrate ($Na_2MoO_4 \cdot 2H_2O$) in silica-free water and dilute to one liter (Note 4).

Monochloroacetic Acid. Dissolve 100 g of reagent-grade acid in silica-free water and dilute to one liter (Note 4).

Reduction Reagent. Dissolve 20 g of pure metol (Note 5), 12 g of sodium sulfite heptahydrate ($Na_2SO_3 \cdot 7H_2O$), and 10 g of sodium oxalate ($Na_2C_2O_4$) in silica-free water and dilute to one liter. The solution should be stable for at least a month, but must be stored in the dark.

Artificial Sea Water. Dissolve 24.7 g of sodium chloride, 13.0 g of $MgC\ell_2 \cdot 6H_2O$, and 9.0 g of $Na_2SO_4 \cdot 10H_2O$ in 954 mℓ of silica-free water. Store in a plastic bottle.

(b) Treatment

Preparation of the Samples. The analyses should be made as soon as possible after collection. Samples should be collected in polyethylene bottles. If delay is unavoidable, store the bottles in the dark to prevent either solution of particulate silica or consumption of silica by diatoms.

The "Yellow" Procedure. Filter samples through a membrane

filter (Millipore, 1 μ). Pipet 50 mℓ of filtered sample into two 100-mℓ polyethylene bottles. To each, add 1 mℓ of monochloroacetic acid solution. Add 1 mℓ of molybdate reagent to the first (sample) bottle, but not to the second (reference) bottle. To the reference bottle, add 1 mℓ of artificial sea water.

Shake the samples, and store for six hours (up to three days). Transfer the contents to a photometer cell or cuvette (Note 6) and measure the absorbance of the sample and the reference solutions at 390 mμ.

The "Blue" Procedure. If low concentrations of silicate are expected or found, continue the procedure by adding 5 mℓ of reduction reagent to each of the 100-mℓ bottles. After eight hours, measure the absorbance of the sample and reference solutions at 640 mμ (Note 7).

The Reagent Blank. This is determined by following either the "yellow" or the "blue" procedure and using 50 mℓ of silica-free water. Use distilled water in the reference cell. The reagent blank should be determined in triplicate for each batch of reagents. If the absorbance of the reagent blank exceeds 0.05 in 5-cm cells, the reagents should be changed or re-purified (Note 8). Also be sure to test the cells in a cell-to-cell blank.

Calibration. Dilute 10 mℓ of standard silica solution to 100 mℓ in a volumetric flask, using silica-free water. Pipet 2 mℓ of this solution (or 6 mℓ for the "yellow" procedure) into a 100-mℓ volumetric flask and dilute to the mark with silica-free artificial sea water. Transfer 50 mℓ of the seawater solution to a 100-mℓ polyethylene bottle and follow either the "yellow" or the "blue" procedure. Use 50 mℓ of silica-free distilled water for the reference and add the reagents as directed. A blank correction is not needed, but the calibration should be made in triplicate (Note 9).

(c) Interference

Grasshoff[3] reports that phosphates do not interfere in the α-silicomolybdic acid ("yellow") procedure in concentrations up to 4 μg-at PO_4-P/liter. The interference with the "blue" procedure is more serious, but this can be avoided by addition of oxalate ions to the sample. In this case, there is no interference by phosphate at concentrations up to 2 μg-at PO_4P/liter.

NOTES

1. All reagents should be stored in hard polyethylene bottles. Bottles made of normal high-pressure polyethylene are permeable to water vapor and should not be used for storage of standard solutions for long periods of time.
2. A slight excess over the theoretical amount is used following the suggestion of Strickland and Parsons.[4]
3. All glassware should be allowed to stand overnight in a 1:1 mixture of sulfuric and nitric acids to make the

glassware insoluble. The glassware must be free of organic solvents, particularly acetone, before this treatment. After the treatment, wash the glassware thoroughly with tap, distilled, and silica-free water. The glassware should not be allowed to dry before use.

4. The reagent is stable indefinitely.
5. Metol is p-methylaminophenol sulfate.
6. Cells may be 1, 2, 3, up to 5 cm, depending on the concentration of silica.
7. Storage of the samples for up to 60 hours has little or no effect on the absorbance.
8. The silica content of the reagents may be a significant fraction of the total silica content when the "blue" procedure is applied to the sample.
9. Only one calibration point is needed because the absorbance is exactly a linear function of the silica concentration.[3]

REFERENCES

1. K. Grasshoff and H. Hahn, Z. Anal. Chem., 168, 247 (1959).
2. J. D. H. Strickland, J. Am. Chem. Soc., 74, 862, 868, 872 (1952).
3. K. Grasshoff, Deep-Sea Res., 11, 597 (1964).
4. J. D. H. Strickland and T. R. Parsons, A Practical Handbook of Seawater Analysis, Fisheries Research Board of Canada, Bulletin No. 167, 1968, p. 69.

10

ORTHOPHOSPHATE

10.1 Introduction

Phosphorus occurs in natural waters in a variety of forms, though it is possible to separate these into convenient, if arbitrary categories. These include inorganic phosphorus, dissolved and particulate; and organic phosphorus, dissolved and particulate. Certain definitions have been made that permit several fractions to be determined. Empirical distinctions can be made on the basis of particle size, reactivity with acidified molybdate, the ability of fuming perchloric acid to convert any form of phosphorus to orthophosphate ion, and the evident ability of ultraviolet light to convert organic phosphate to orthophosphate without affecting polyphosphate, which can be hydrolyzed with perchloric acid. By means of these distinctions, it is possible to distinguish up to eight phosphorus fractions, but some fractions may lack convenient methods of analysis or may not be of especial significance ecologically or even be present at any given time.[1] Most procedures do depend on conversion of some phosphorus fraction to orthophosphate ion.

Currently, most commonly used sensitive methods for the estimation of orthophosphate depend upon the reaction of orthophosphate with an acidified molybdate solution to form a phosphomolybdate heteropolyacid (Note 1). Typically, the phosphomolybdate is reduced to an intensely colored phosphomolybdenum blue, and solutions obey Beer's Law under suitable conditions, up to concentrations of at least 90 μg PO_4-P/liter. When the phosphate concentration is less than 0.03 μg-at PO_4-P/liter, the direct determination is impractical using 10-cm cells. The sensitivity of the method can be increased by the Proctor-Hood procedure,[2] which consists in extracting the phosphomolybdenum blue into isobutanol or into _n_-hexanol.

The determination of orthophosphate by the molybdenum blue method depends upon working conditions which have been developed on an empirical optimum basis. At least five variations of the original Denigés method have been reported, and four major procedures have been used during the past decade including those of Harvey,[3] Wooster and Rakestraw,[4] Murphy and Riley,[5] and Proctor and Hood.[2] These methods have been compared by Jones and Spencer.[6] The first two use stannous chloride as a reducing agent, the third uses ascorbic acid, and the last uses an extraction technique. Several parameters are involved and these may be compared in a general way.

Temperature. For most procedures, at a given concentration of orthophosphate, temperature increase results in a decrease in the time required for maximum color development, an increase in the observed intensity, and a reduction in the time the color is stable. The time interval between reduction and measurement is critical. The Murphy-Riley procedure seems to be exceptional, and no significant temperature coefficient has been observed between 15 and 30°C.[5] Also, time between reduction and measurement seems to be less critical with this procedure; the color seems to be stable for 24 hours, though this is surely dependent upon the nature of the phosphorus compounds present.[5]

Acidity and Molybdate Concentration. Color may develop in the absence of phosphate when the acid:molybdate ratio used is too high or too low; thus, reagent blanks must be obtained. Many workers prefer the higher reagent concentrations recommended by Harvey,[3] i.e., 0.05% ammonium molybdate and 0.28 N acid.

Reducing Agent Concentrations. With stannous chloride, the intensity of color is dependent upon the tin-phosphorus ratio up to about 13; at higher ratios, the effect is smaller. High stannous concentrations increase the tendency for green tints to be formed, and this requires selection where the interference is minimal (700-705 mμ).[4]

Stability of Reducing Agent. Stannous chloride is readily oxidized upon exposure to air, and many workers prefer to make fresh solutions before each set of analyses; others have found that the reagent may be stored under mineral oil or petroleum ether and in contact with a piece of mossy tin. Ascorbic acid also is easily oxidized, and solutions should be prepared on the day they are used. Stabilization by the addition of ethylenediamine tetraacetic acid (disodium salt) and formic acid has been suggested,[5] though Strickland and Parsons[1] note that the reagent is stable for many months if frozen in a plastic bottle.

Salt Error. This is appreciable with the stannous chloride methods, and the procedure is calibrated using artificial sea water.[3,4] The modified ascorbic acid procedure has no appreciable salt error (less than 1%) and phosphate-free distilled water can be used for calibrations.[5]

Interference from Metals.[6] Copper interferes seriously with the stannous chloride procedures at concentrations greater than 50 μg/liter. Also, arsenite-ion in sea water (~2.5 μg AsO_3-As/liter) produces "high" results with the stannous chloride procedure, though the interference seems negligible with the Murphy-Riley procedure.

Hydrolysis of Phosphorus Compounds. The possibility of hydrolysis of organic phosphorus compounds under the acidic conditions used is ever present. The positive error that hydrolysis represents can be minimized by short reaction time, e.g., 10 minutes, and it appears that with the Murphy-Riley procedure, the error is significant only for very labile compounds,[5] such as might be found in cultures, in nearshore waters containing more organic matter than is typical of oceanic waters, or in waters where the biological activity is especially high.[6]

In summary, it appears that all methods have some limitations, that some show a definite seasonal discrepancy (because of the presence of arsenite or labile phosphorus compounds or some interfering species related to biological activity), and that there is little difference in the repeatibility of all methods in the hands of careful workers. The method of personal choice is possibly to be decided in terms of convenience and personal experience, and for these reasons, the modified ascorbic acid procedure is recommended here for intermediate phosphate levels; for low phosphate levels, the Stephens[7] modification of the Proctor-Hood procedure is recommended. If serious interference from arsenic is anticipated (certain fresh-water or estuarine samples), an alternative extraction procedure should be considered that is reported to be less sensitive to arsenic interference.[10]

10.2 Determination of Orthophosphate (moderate levels)

The modified ascorbic acid procedure of Murphy and Riley[5] consists in treating phosphate-containing sample with a single reagent, which is a mixture of acidified molybdate, ascorbic acid, and antimony. The blue-purple reduction product contains antimony and phosphorus in a 1:1 atomic ratio. The antimony decreases the time needed for formation of the colored product, and reduces the possibility of hydrolysis. The only limitation of the procedure, the need to prepare "mixed reagent solution" each day of use, is a minor one.

The procedure given here can be used to determine orthophosphate in the range of 0.03 to at least 25 μg-at PO_4-P/liter. The limit of detection corresponds to the lower value (using a

5-cm or 10-cm cell), and the upper value corresponds to an absorbance of about 0.65 (2-cm cell).

Procedure

(a) Reagents

Sulfuric Acid (5 N). Cautiously pour 280 mℓ of concentrated acid (96%, sp gr 1.84) into about 500 mℓ of triple-distilled water (Note 2) and dilute to one liter.

Ammonium Molybdate. Dissolve 40 g of reagent-grade ammonium molybdate, $(NH_4)_6Mo_7O_{24} \cdot 4H_2O$, in about 900 mℓ of triple-distilled water and dilute to one liter. Store in a polypropylene bottle out of the light or in an amber, phosphate-free (see cleaning procedure) borosilicate glass bottle. The reagent should be stable for about a year.

Potassium Antimonyl Tartrate. Dissolve 1.3615 g of reagent-grade material (tartar emetic, $K(SbO)C_4H_4O_6 \cdot 1/2H_2O$) in triple-distilled water and dilute to 500 mℓ. The solution is stable for several months when stored in a glass bottle.

L-Ascorbic Acid. Dissolve 1.32 g of reagent-grade material in 75 mℓ of triple-distilled water.

Mixed Reagent Solution. Add 145 mℓ of sulfuric acid (5 N) to a 250 mℓ graduated cylinder and add 37.5 mℓ of ammonium molybdate solution. Mix thoroughly. Add ascorbic acid (75 mℓ) and again mix thoroughly. Finally, add 12.5 mℓ of potassium antimonyl tartrate solution and mix thoroughly. The mixed reagent solution is stable only about 24 hours and should be prepared as needed.

Stock Phosphate Solution (2.5 g-at PO_4-P/mℓ). Dissolve 0.3400 g reagent-grade anhydrous (Note 3) potassium dihydrogen phosphate (KH_2PO_4) in distilled water and dilute to one liter. Stored in a phosphate-free amber bottle with 1 mℓ of chloroform, the solution should be stable for several months.

(b) Cleaning Procedure

A preliminary treatment of glassware with phosphate reagents should be made to remove phosphate that may be adsorbed on the surfaces from laboratory reagents. Once the glassware has been properly cleaned, it should be reserved for phosphate determinations and should not be cleaned with laboratory detergents.

The following steps are involved.

Clean and thoroughly rinse all glassware in the usual manner.

Add 8 mℓ of mixed reagents for each 100 mℓ of volume, add distilled water, and mix thoroughly. Be sure that all internal surfaces are in contact with the mixture.

Allow the mixture to stand for 20 minutes. Discard the mixture and rinse each container thoroughly with triple-distilled water. Once the cleaning procedure has been performed, the glassware can be reused without repeating the treatment, unless detergent has been accidentally used. Store the glassware in 0.1% sulfuric acid solution (1 g of concentrated acid per liter of distilled water).

(c) Storage of Samples

Samples intended for phosphate analyses should not be stored in polyethylene bottles at room temperature, in view of available evidence[8,9] on the loss of phosphate to these containers. Samples should be analyzed within one-half hour if possible, or at a minimum within two hours; otherwise, some provision for proper storage must be made. Three procedures are available.

1. Preservation. Murphy and Riley[5] recommend filtration of the sample and preservation by adding chloroform (1 mℓ per 150 mℓ of sample). Some workers report the inefficiency of this procedure, at least in their hands.

2. Bottle Treatment. Heron[8] recommends filling the polyethylene bottles with a solution of iodine-potassium iodide (5 g of iodine and 8 g of potassium iodide per 100 mℓ of distilled water). After standing for one week, the bottles are drained and washed well with distilled water. When treated bottles are used, phosphate losses are reported to be minimal, but a modification in the Proctor-Hood phosphate method must be made. Others[9] recommend even glass bottles be treated to prevent uptake of phosphate.

3. Freezing. Samples in polypropylene bottles are quick-frozen in a Dry Ice-acetone bath, then stored in a deep-freeze unit at the lowest possible temperature. This is reported[1] to stabilize samples for many months, even those drawn from euphotic zones of subtropical and tropical waters.

(d) Analysis

The following steps are involved:

1. Filter the sample, warmed to room temperature, using a 0.45 μ membrane filter.

2. Add 40 mℓ of sample to a 50-mℓ graduated mixing cylinder.

3. Add 8 mℓ of mixed reagents and dilute to 50 mℓ. Mix thoroughly.

4. After ten minutes, and within three hours, read the absorbance of the solution in a 2- or 5-cm cell at 690 mμ. Use distilled water in the reference cell.

5. Cell-to-cell blank should be checked. The absorbance should be 0.000 when both sample and reference blanks are filled with distilled water. If not, correction can be made along with reagent correction.

6. Reagent blanks should be run periodically. The analysis procedure is followed (steps 2-5) except that 40 mℓ of triple-distilled water is used instead of sample. The absorbance value should be less than 0.005. If not, the distilled water should be checked first, then the molybdate reagent.

(e) Calibration Curve and Calculations

Prepare a working standard solution by diluting 10.0 mℓ of stock phosphate solution to one liter with triple-distilled water, and store in a dark bottle with 1 mℓ of chloroform. The solution should not be used after standing for two weeks.

A calibration plot can be prepared by taking n milliliters of working standard, diluting to 40 mℓ in a 50-mℓ graduated mixing cylinder and carrying out the analysis. The concentration is equal to 0.625n μg-at PO_4-P/liter, where n is the number of milliliters used. Suggested values of n are 0.00, 0.20, 0.50, 0.70, 1.0, 3.0, 7.0, 10.0.

The concentration of an unknown sample can be obtained from the calibration plot, μg-at PO_4-P/liter as a function of absorbance corrected for reagent blank, or from the corresponding linear relationship

$$\mu\text{g-at } PO_4\text{-P/liter} = m(A_s - A_b) \qquad (10\text{-}1)$$

Here, m is the slope of the linear relationship and is equal to 38.0 using a 2-cm cell, and is proportionately less for longer cells; A_s and A_b are the absorbances of the sample and reagent blanks, respectively.

An alternate approach can be used. The average values of A_s and A_b are determined using four replicate samples. To determine A_s, dilute a 5-mℓ sample of the working standard to 40 mℓ; use distilled water to determine A_b. The two values should not be corrected for cell-to-cell blanks. The concentration of the unknown sample is calculated from

$$\mu\text{g-at } PO_4\text{-P/liter} = A_u C \qquad (10\text{-}2)$$

Here, A_u is the absorbance of the unknown, and C is a conversion factor that is equal to $3/(A_s - A_b)$.

A salt correction is evidently not needed with either procedure. Also, the calibration plot or the value of C should remain constant provided fresh reagents are used, and the same glassware and spectrophotometer are used. Nevertheless, the value of C or a point on the calibration curve should be checked periodically.

10.3 Determination of Orthophosphate at Low Levels

This procedure is a combination of the extraction method and the mixed reagent method used in the preceding section. The effective concentration range that can be readily determined (~0.006-0.30 μg-at PO_4-P/liter) make this procedure equally useful for many fresh waters and oligotrophic marine waters. The limit of detection is about 0.006 μg-at PO_4-P/liter. The procedure is essentially that described by Stephens.[7]

Procedure

(a) Reagents

The following reagents are needed in addition to those listed in Section 10.2.

Isobutanol. If analytical-quality, reagent-grade material is not available, purify isobutanol by the following procedure. Extract impurities in 500 mℓ of alcohol with three 50-mℓ aliquots of water. Dry the alcohol over anhydrous sodium carbonate, filter the alcohol into a distillation apparatus, and collect that fraction boiling at 107-109°C and 760 mm.

Absolute Ethanol. Use reagent-grade material.

(b) Cleaning Procedure

See Section 10.2.

(c) Storage of Samples

See Section 10.2

(d) Analysis

To a 200-mℓ sample in a 250 mℓ separatory funnel, add 20 mℓ of mixed reagents, and mix thoroughly. Wait ten minutes, then add 29 mℓ of purified isobutanol and shake for exactly one minute. Allow the contents of the separatory funnel to stand for five minutes, then separate the alcohol extract. Dilute to exactly 10 mℓ with absolute ethanol, pour into a dry 10-mℓ, 10-cm cell, and measure the absorbance at 690 mμ with the reference cell filled with isobutanol.

(e) Calibration and Calculations

Correct the absorbance using a reagent blank and a cell-to-cell blank (isobutanol in both reference and sample cells).

Prepare the working standard solution described in Section 10.2. Prepare either a calibration plot or use the increment method of calibration.

A calibration plot is prepared by taking n milliliters of working standard, diluting to 200 mℓ, and carrying out the analysis. The concentration is equal to 1.25×10^{-3} n μg-at PO_4-P/liter, where n is the number of milliliters of working standard solution used. Suggested values of n are 0.0, 4.0, 10, 15, 20, 25, 30. The concentration of an unknown sample is obtained from the calibration plot.

If the increment method of calibration is used, determine average values of A_s and A_b (see Section 10.2) using four replicate samples. To determine the value of A_s, dilute a 15-mℓ sample of the working standard to 200 mℓ; 200 mℓ of triple-distilled water is used to determine A_b. Do not correct for cell-to-cell blanks. The concentration of the unknown sample is calculated from

$$\mu\text{g-at } PO_4\text{-P/liter} = A_u C \qquad (10\text{-}3)$$

Here A_u is the corrected absorbance of the unknown sample, and C is a conversion factor that is equal to the reciprocal of the calibration plot slope. Analyses may be carried out on fresh and marine waters.

NOTES

1. Sensitive spectrofluorimetric methods have also been reported.[11] One is based on formation of an ion-association with molybdophosphate with Rhodamine B. Advantages of sensitivity and selectivity are described.
2. Alternatively, see Section 8.4.
3. Dry in an oven at 105° for three hours.

REFERENCES

1. J. D. H. Strickland and T. R. Parsons, A Practical Handbook of Seawater Analysis, Fisheries Research Board of Canada, Bulletin No. 167, 1968.

2. C. H. Proctor and D. W. Hood, J. Mar. Res., 13, 122-132 (1954).

3. H. W. Harvey, The Chemistry and Fertility of Sea Water, Cambridge University Press, 1955.

4. W. S. Wooster and N. W. Rakestraw, J. Mar. Res., 10, 91 (1951).

5. J. Murphy and J. P. Riley, Anal. Chim. Acta, 27, 31 (1962).

6. P. G. Jones and C. P. Spenser, J. Mar. Biol. Assoc. U.K., 43, 251 (1963).

7. K. Stephens, Limnol. Oceanogr., 8, 361 (1963).

8. J. Heron, Limnol. Oceanogr., 7, 316 (1962).

9. W. Hassen-Teufel, R. Jagitsch, and F. F. Koczy, Limnol. Oceanogr., 8, 152 (1963).

10. W. Chamberlain and J. Shapiro, Limnol. Oceanogr., 14, 921 (1969).

11. G. F. Kirkbright, N. Narayanaswamy, and T. S. West, Anal. Chem., 43, 1434 (1971).

11

TOTAL PHOSPHORUS

11.1 Introduction

As noted in Chapter 10, phosphorus may exist in the form of inorganic or organic phosphorus compounds and these compounds may be soluble or present in particulate matter. Changes in phosphorus content have been correlated with water movements and plant growth and general productivity. Much of the information about the distribution of phosphorus has been concerned with inorganic phosphorus, which is largely present as orthophosphate ions ($H_2PO_4^-$ and HPO_4^{2-}).[1,2] Comparatively little information is available concerning the distribution of organic phosphorus, probably because of the difficulties various workers have experienced in its determination.

The concentration of dissolved organic phosphorus may be sizable in certain areas (in excess of 1 μg-at P/liter in the N.E. Pacific Ocean, for example[3]), but the nature of dissolved organic phosphorus is unknown. Presumably, as Armstrong[1] notes, organic phosphorus compounds include phosphoproteins, nucleoproteins, sugar phosphates, and their oxidation products.

The major difficulties in the determination of organic phosphorus center around the need to convert the organic phosphate to orthophosphate ion by oxidizing the organic material under acid conditions. Orthophosphate is then determined as usual by the modified Denigé's method (Chapter 10). In effect, the extra step introduces the possibility of contamination because of the extra reagents or manipulations required. Not only must the oxidation be complete, but the oxidizing agent must be destroyed and excess acid must be destroyed before continuing with the analysis of orthophosphate ion.

Two methods of destroying the organic material have been used: (1) use of chemical oxidizing agents (hydrogen peroxide, sulfuric acid, perchloric acid, peroxydisulfate ion) and (2) photochemical oxidation. Both methods are presented here.

11.2 Chemical Oxidation Procedure

The procedure used here is based on a method developed by Hansen and Robinson.[4] The organic material is oxidized with perchloric acid, which seems to be superior to sulfuric acid. Total phosphorus is determined and organic phosphorus represents the difference between total and inorganic phosphorus.

Procedure

(a) Reagents

In addition to the reagents used for inorganic phosphorus, the following reagents are needed (Note 1).

Perchloric Acid. Use reagent-grade 60% acid (Note 2).

Sodium Hydroxide (1 M). Dissolve 40 g of reagent-grade NaOH in distilled water and dilute to one liter.

Hydrochloric Acid (12 M). Use concentrated (37%, sp. gr. 1.19) acid.

(b) Analysis

Filter or centrifuge the sample (Notes 3,4). Add 3 mℓ of perchloric acid (Note 2) to a 40-mℓ sample in a 125-mℓ Erlenmeyer flask. Evaporate the sample on a hot plate until the fuming temperature of perchloric acid is reached and fumes are noted. Place a watch glass on the flask and heat the contents for five minutes at a temperature just below the fuming temperature.

Remove the flask and contents from the hot plate and cautiously add 3 mℓ of (12 M) hydrochloric acid (concentrated) to the contents of the flask. Replace the flask on the hot plate and fume off rapidly. This treatment should remove any arsenic, which interferes in the subsequent determination.

Allow the flask and contents to cool. Add 30 mℓ of distilled water to dissolve the salts. Neutralize the excess acid by adding the sodium hydroxide solution dropwise until neutral to litmus. Transfer the solution to a 50-mℓ mixing cylinder, dilute to 40 mℓ with distilled water, and continue the analysis using the mixed reagent procedure, Section 10.2(d), step 2.

11.3 Photochemical Oxidation

Ultraviolet photochemical oxidation has been applied to the analysis of sea water,[7,8] algal suspensions,[8] and lake water.[9] Construction details of photochemical reactors that irradiate six[8] or twelve[9] silica tubes simultaneously have been given. Another commercially available apparatus has a superior geometry, but can handle only one sample at a time (Note 5).

The speed of photochemical decomposition depends upon the compound and perhaps upon the salinity. The decomposition rate for phosphorus follows first-order kinetics. Specific first-order rate constants k have been measured for various phosphorus compounds in sea water and in distilled water, and the values are compared in terms of decomposition half-life values $t_{1/2}$ (Table 11-1). Five half-lives would be required to achieve 97% decomposition.

Three phosphorus fractions--orthophosphate, total organic phosphorus, and organic polyphosphate--can be determined with this procedure. The operational definitions are given in Table 11-2.

TABLE 11-1

Half-Life Values for Photochemical Decomposition of Phosphorus Compounds Using a 380-W Mercury Arc Lamp[8]

	Half-life $t_{1/2}$, hr[b]	
Compound[a]	Distilled water	Sea water
Glucose 6-phosphate	0.0186	0.159
Adenosine monophosphate	0.140	0.126
Triphenyl phosphine	0.578	0.381
Sodium pyrophosphate	—	28.57

[a] 10 μ M solutions, maximum temperature 60°C.

[b] $t_{1/2} = 0.693/k$.

TABLE 11-2

Phosphorus Fractions in Sea Water: Operational Definitions and Typical Concentrations

Species	Operations[a]	Concentration, μg-at P/liter	
		Defined value[b]	Typical value[c]
Orthophosphate	A	a	~40
Total organic	A+B+C	c - a	0.2-0.3
Organic polyphosphate	A+B+C	c - b	0.02-0.03

[a]Operations: A, determination of initial orthophosphate; B, irradiation; C, acid hydrolysis following irradiation.

[b]Concentrations are those observed after each operation.

[c]English Channel, April 1967.[8]

Procedure

(a) Apparatus

The choice of apparatus depends upon the number of samples to be determined simultaneously. The procedure used here is based on a six-sample unit that uses a 380-W mercury arc lamp.[8] A single sample unit is available (Note 5), but different irradiation times are involved; see Section 11.2(d).

(b) Reagents

Hydrogen Peroxide. Use 30% reagent-grade material. Avoid contact with the skin.

Hydrochloric Acid. Use 12 M acid (37%, sp. gr. 1.19).

Calibration Solution. Weigh 0.0655 g of triphenyl phosphine in a silica reactor tube, add two drops of conc. HCℓ, and dilute to 100 mℓ, (1 mℓ = 2.5 μg-at P).

(c) Analysis

First determine the initial orthophosphate concentration (Section 10.2). Let this concentration be equal to a μg-at P/liter (cf. Table 11-2).

Next treat a 100-mℓ sample with two drops of hydrogen peroxide and irradiate for five hours. To determine the organic polyphosphate concentration, add phosphorus-free distilled water and dilute to 100 mℓ before determining the orthophosphate concentration (= b μg-at P/liter).

To determine the total organic phosphorus, treat 100 mℓ of the irradiated sample with 0.2 mℓ of conc. hydrochloric acid and boil for one hour. When cool, dilute to the original volume (with allowance for sample removed). Determine the orthophosphate concentration (= c μg-at P/liter).

The concentrations of various phosphorus fractions are defined in terms of the operations given in Table 11-2.

(d) Calibration

The unit should be checked for irradiation efficiency with 100 mℓ of calibration solution. Samples can be removed at selected intervals and tested for orthophosphate concentration. About three hours of irradiation should produce complete conversion with the 380-W tube (or five hours for most samples). A reduction to one hour is possible with a 1200-W tube.

NOTES

1. All glassware should be specially cleaned following directions given in Chapter 10.
2. Concentrated perchloric acid must be handled with care[5] and must not be allowed to come in contact with organic compounds (wood, alcohol, skin, or material such as plankton, mud, or combustible material). Contact with combustible material must be avoided. It must be stored under fireproof conditions (because of the potential danger of formation of anhydrous perchloric acid in the event of a fire).
3. Samples of plankton should be treated with small portions of nitric acid (and warmed) before being treated with perchloric acid.
4. Some organic phosphorus compounds show a remarkable resistance to decomposition by perchloric acid, and a repeated treatment may be necessary for a particular sample.[6]
5. The combustion apparatus was used by Manny and co-workers[9] and is available (Model 6515) from Ace Glass, Inc., Vineland, N. J. These workers used a 450-W mercury arc lamp (679A, Engelhard Hanovia, Inc., Newark, N. J.).

REFERENCES

1. F. A. J. Armstrong, Oceanogr. Mar. Biol. Ann. Rev., 3, 79 (1965).

2. J. P. Riley, in Chemical Oceanography (J. P. Riley and G. Skirrow, eds.), Vol. 2, Academic Press, New York, 1965, Chapter 21.

3. J. D. H. Strickland and K. H. Austin, J. Fish. Res. Bd. Can., 17, 337 (1960).

4. A. D. Hansen and R. J. Robinson, 12, 31 (1953).

5. J. H. Kuney, Chem. Eng. News, 25, 1658 (1947).

6. J. D. Burton and J. P. Riley, Mikrochim. Acta, 1350 (1958).

7. F. A. J. Armstrong, P. M. Williams, and J. D. H. Strickland, Nature, 211, 481 (1966).

8. F. A. J. Armstrong and S. Tibbitts, J. Mar. Biol. Assoc. U.K., 48, 143 (1968).

9. B. A. Manny, M. C. Miller, and R. G. Wetzel, Limnol. Oceanog., 16, 71 (1971).

12

INORGANIC NITROGEN COMPOUNDS—I. AMMONIA

12.1 Introduction

Until recently, routine determinations of ammonia were rarely made aboard ship, probably because of the lack of a rapid and sensitive method. When available methods have been used, large variations among replicate samples have been observed, and Cooper[1] has attributed this to adsorption of ammonium ions by particulate matter. There was also the danger that the alkaline conditions which were used caused breakdown of combined nitrogen and liberated ammonia. Finally, in dealing with polluted waters, it is possible that organic nitrogen and urea in septic waters will be subject to biological action, with the progressive formation of ammonia.

The classical methods for determining ammonia involve Nesslerization and/or distillation.[2]

Nesslerization. Nessler's reagent is a strongly alkaline solution of potassium tetraiodomercurate(II), K_2HgI_4, which combines with ammonia to form a yellowish brown colloidal dispersion:

$$2K_2HgI_4 + NH_3 + 3KOH \rightarrow I\text{-}Hg\text{-}O\text{-}Hg\text{-}NH_2 + 2H_2O \qquad (12\text{-}1)$$

The intensity of the color follows Beer's Law at 425 mμ, but there are several interferences.

Nessler's reagent forms a precipitate with several ions,

including calcium, magnesium, iron, and sulfide. In an attempt to obviate this interference, the ammonia was distilled out and then treated with Nessler's reagent, but this procedure gives erroneously high results because of the breakdown of combined nitrogen under the conditions used. Other workers have used Rochelle salt solution (Wattenberg method) or chelating agents to prevent precipitation with calcium or magnesium.[2] Calcium and magnesium have also been removed by preliminary precipitation,[3] although the method tends to be time-consuming.

Certain organic compounds in polluted waters tend to produce an off color with Nessler's reagent. This problem can be obviated by distillation, if the organic materials are non-volatile.

The natural sample may contain a turbidity and a color which interferes in the colorimetric determination. Flocculation may eliminate the problem; if not, it may be necessary to use the distillation method.

Distillation. The water sample is made alkaline and the ammonia is volatilized, collected in the distillate, and absorbed. Riley prevented decomposition of organic nitrogen compounds by carrying out the distillation at a pH of 9.2 (metaborate buffer) under reduced pressure in a stream of air. Ammonia in the distillate could be determined by conventional Nesslerization, but Riley[4,5] considers determination by an indophenol blue method to be preferable. Subsequent modifications have eliminated the troublesome distillation of ammonia (Section 12.3).

Another popular approach, the pyrazolone method (Section 12.2), has a disadvantage of an extraction step that is balanced by the sensitivity of the procedure.

12.2 Pyrazolone Method

This method consists in treating a sample with chloramine T, followed by an aqueous solution of 3-methyl-1-phenyl-5-pyrazolone and bis(3-methyl-1-phenyl-5-pyrazolone), and extracting the yellow complex formed with ammonia into carbon tetrachloride. The absorbance of the carbon tetrachloride solution is measured. This method was devised to measure ammonia in industrial wastes,[6] and has been applied to the analysis of ammonia in sea water.[7] The method obviates the need for a distillation procedure and avoids the possibility of contamination.

Procházhová[8] has studied this complicated reaction and has suggested a reasonable mechanism for the conversion of ammonia to rubazoic acid. The procedure here is a modification[7,8] that eliminates the objectionable use of carbon tetrachloride as an extraction agent.

Procedure

(a) Reagents (Note 1)

Ammonia-Free Water. This water must be used to prepare all reagents and stock solutions. Ammonia-free water may be prepared by (1) shaking distilled water with Folin's ammonia permutit, (2) passing the water through a small column containing exchange resin, such as Amberlite IR120H or Zeo-Carb 215 (see Section 8.4), (3) redistilling water that has been treated with bromine and allowed to stand overnight, or (4) redistilling water that has been treated with 1.5 g of $KMnO_4$ and 0.1 mℓ of concentrated sulfuric acid per liter. In methods 3 and 4 collect only the middle fraction while passing ammonia-free air into the receiving flask.

Artificial Sea Water (Note 2).

Trichloroethylene. Use reagent-grade material.

Sodium Carbonate (0.5 N). Dissolve 53 g of Na_2CO_3 (or 143 g of $Na_2CO_3 \cdot 10H_2O$) in ammonia-free distilled water and dilute to one liter.

Hydrochloric Acid (0.5 M). Pour 10 mℓ of conc. HCℓ (37%) into ammonia-free distilled water and dilute to 240 mℓ.

Sodium Hydroxide (1 M). Dissolve 4 g of NaOH in ammonia-free distilled water and dilute to 100 mℓ. Store in a plastic wash or dropper bottle.

Citrate-Phosphate Buffer. Mix 0.1 M citric acid (79.1 mℓ, Note 3) and 0.2 M Na_2HPO_4 (120.9 mℓ, Note 3). The pH should be 5.8. Stabilization with a few drops of 1:1 toluene-carbon tetrachloride is recommended.[8]

Monoreagent, 3-Methyl-1-Phenyl-5-Pyrazolone, Solution. Strickland and Austin[7] recommend a single recrystallization from hot water. Dissolve 1.25 g in ammonia-free hot water and dilute to 500 mℓ. The reagent is stable indefinitely.[7]

Bispyrazolone (Note 4). Dissolve 0.2 g of bispyrazolone in 40 mℓ of 0.5 N Na_2CO_3 at 90°. After cooling to room temperature, add 80 mℓ of 0.5 N Na_2CO_3. The solution should be freshly prepared for each batch of determinations, though it is reportedly stable for 10 days if stored at 0° under nitrogen.

Standard Ammonium Salt Solutions. Solution A. Dissolve 3.819 g of ammonium chloride in a liter of ammonia-free water.

Solution B (10 μg N/mℓ). Dilute 10 mℓ of Solution A to one liter with ammonia-free water.

Solution C (0.5 μg N/mℓ). Dilute 50 mℓ of Solution B to one liter with ammonia-free water.

(b) Analysis

The first step of the two-step color formation is especially temperature-dependent, and the reaction mixture should be maintained at 15°. Prior to extraction, the reaction mixture should be kept at 15-20°.

Adjust the pH of a 50-mℓ sample of natural water to 6-7 with dropwise addition of 1 M NaOH. Add 5 mℓ of citrate-phosphate buffer, then 2 mℓ of chloroamine-T solution. Mix after each addition. Let the mixture stand for exactly five minutes at 15°. Then, shaking the solution, quickly add 5 mℓ of bispyrazolone solution. After five minutes, add 10 mℓ of pyrazolone solution. Wait until a blue color is formed, if it disappears, then acidify the mixture with 2 mℓ of 0.5 M HCℓ.

Extract with one 10-mℓ portion of trichloroethylene,

allowing equilibration for three minutes. Separate the organic phase through a dry filter plug into a dry, 1-cm spectrophotometer cell and measure the absorbance at 450 mμ using distilled water in the reference cell.

Repeat the determination using ammonia-free synthetic sea water to determine the reagent blank (A_b = 0.040).

(c) Calibration

Salts in sea water suppress the absorbance for a given amount of ammonia, and at a salinity of about 30 parts per thousand, the absorbance is about 60% of the value in pure water. The absorbance-salinity relationship is curved, but is approximately linear in the salinity range of 25-35 parts per thousand.

For analysis of sea water, prepare the calibration solutions by diluting x milliliters (Note 5) of Solution C to 100 mℓ with artificial sea water and following the analysis procedure. The concentration of the unknown solution may be determined by reference to the calibration plot (corrected absorbance versus μg N/liter).

For analysis of fresh water, prepare the calibration solutions by following the same procedure, using distilled water instead of artificial sea water. For an ammonia concentration range of 0.025-2.5 ppm, a minor modification is necessary (Note 6).

12.3 Phenol-Hypochlorite Method

This method is one of many modifications of the

indophenol reaction.[7,9-12] At conditions of high pH, ammonia, phenol, and hypochlorite ion combine to produce a blue color due to indophenol. Generally, limitations of various procedures have been due to two complications: interference by calcium and magnesium ions in sea water, or lack of specificity. The first problem either results in insensitivity or requires considerable attention to detail.

The method described here is based on one developed by Solórzano[12] and has several advantages. The method is reasonably rapid and uncomplicated; distillation and extraction are obviated. No interference by representative amino acids and urea was noted. The procedure, admittedly, has a lower sensitivity than a similar one of Richards and Kletsch[11] that measures a considerable fraction of the amino acid nitrogen (~80%) as well as smaller fractions of hydroxyl amine (~50%) and urea nitrogen (~5%).[13] Measurement of the other forms of trivalent nitrogen was thought to be an advantage, assuming the results were to be used in studies of productivity. This presupposes that the method measures the tervalent nitrogen available to phytoplankters, and presently there is no reason to believe that this is the case.

Procedure

(a) Reagents

All reagents should be prepared using ammonia-free distilled water.

Alcoholic Phenol. Dissolve 10 g of reagent-grade phenol

(avoid contact with skin) in 100 mℓ 95% ethanol.

Sodium Nitroprusside. Dissolve 1.0 g of sodium nitroprusside in 200 mℓ of distilled water, and store in an amber bottle. The solution should not be stored more than one month.

Hypochlorite Solution. Use commercial hypochlorite (e.g., Clorox). The solution should be at least 1.5 N and the strength should be checked (Note 7) periodically because the solution slowly decomposes.

Complexing Solution. Dissolve 100 g of trisodium citrate and 5 g of sodium hydroxide in 500 mℓ of distilled water.

Oxidizing Reagent. Mix 25 mℓ of hypochlorite solution and 100 mℓ of complexing sodium citrate solution. Discard after one day.

(b) Glassware

Initially, wash all glassware with warm dilute HCℓ solution. Rinse very thoroughly with distilled water (ordinary) immediately before using.

(c) Sample Treatment

It seems to be undesirable to store samples, even frozen solid, that will be analyzed for ammonia. Frozen samples may be stable for a few days. Unfrozen samples should be analyzed within a couple of hours. If hydrogen sulfide is present, remove it by acidifying the solution to pH 3 and passing air through the solution until the sulfide odor disappears.

(d) Analysis

The first step consists in adding sequentially the following reagents to a 50-mℓ sample in a glass-stoppered mixing cylinder: phenol solution (2 mℓ), sodium nitroprusside solution (2 mℓ), oxidizing solution (5 mℓ). Mix thoroughly after each addition. Next, allow the color to develop for one hour at room temperature (24 ± 3°). Measure the absorbance at 640 mμ using a 10-cm cuvette.

(e) Calibration and Calculations

Prepare a calibration curve, absorbance corrected for blank absorbance versus concentration of calibration solutions (Note 5). Beer's Law should be observed in the range 0.1-10 μg-at NH_3-N/liter. Determine the conversion factor F and calculate the concentration of natural water sample from the absorbance of the final solution A_u corrected for absorbance of the blank A_b using the relation

$$\text{Concentration } (\mu\text{g-at } NH_3\text{-N/liter}) = (A_u - A_b)F \qquad (12\text{-}2)$$

For this system, the value of F should be about 6.5.

NOTES

1. In all procedures for determining ammonia in sea water, precautions must be taken to avoid contamination by ammonia in the air, by ammonium compounds, and by reagents that may contain ammonium compounds as impurities. Some even have recommended working in a room in which

ammonium compounds are never used.

2. See Chapter 4 for directions. This water should be prepared with ammonia-free water and should be free of ammonia. This may be checked by running a blank. If necessary, the solution can be freed of ammonia by partial evaporation and redilution to the original volume with ammonia-free water.

3. Citric acid (0.1 M): Dissolve 9.6 g of reagent-grade material in ammonia-free water and dilute to 500 mℓ. Na_2HPO_4 (0.2 M): Dissolve 14.2 g of anhydrous material in ammonia-free distilled water and dilute to 500 mℓ.

4. Bis reagent is bis(3-methyl-1-phenyl-5-pyrazolone) or 3,3'-dimethyl-1,1'-diphenyl-[4,4'-bi-2-pyrazoline]-5,5'-dione. This may be purchased (e.g., Eastman Organic Chemicals) or prepared by following the directions of Kruse and Mellon.[6]

5. Suggested values of x are 0.0, 0.5, 1.0, 1.5, 2.0, 2.5, 3.0, 4.0. The concentration will be 5x μg NH_3-N/liter or 0.36x μg-at NH_3-N/liter.

6. Add x milliliters of Solution B and dilute to 100 mℓ with distilled water to obtain concentrations of 0.1x mg NH_3-N/liter or 0.1x ppm.

7. Dissolve about 2 g of potassium iodide in 50 mℓ of distilled water, pipet in 1.0 mℓ of hypochlorite solution, and acidify with five drops of conc. HCℓ. Titrate the liberated iodine with thiosulfate solution (dissolve 1 g

of $Na_2S_2O_3 \cdot 5H_2O$ in 40 mℓ of distilled water). If less than 12 mℓ of thiosulfate are needed to dispel the yellow color, discard the hypochlorite solution.

REFERENCES

1. L. H. N. Cooper, J. Mar. Biol. Assoc. U.K., 27, 322 (1948).
2. Standard Methods for the Examination of Water and Wastewater, 11th ed., American Public Health Association, Inc., New York, 1955, p. 167.
3. H. E. Wirth and R. J. Robinson, Industr. Engng. Chem. (Anal. Ed.), 5, 293 (1933).
4. J. P. Riley, Analyt. Chim. Acta, 9, 575 (1953).
5. J. P. Riley and P. Sinhaseni, J. Mar. Biol. Assoc. U.K., 36, 161 (1957).
6. J. Kruse and M. G. Mellon, Sewage and Ind. Waste, 24, 1098 (1952).
7. J. D. H. Strickland and K. H. Austin, J. Cons. Int. Explor. Mer, 24, 446 (1959).
8. L. Procházková, Anal. Chem., 36, 865 (1964).
9. B. S. Newell, J. Mar. Biol. Assoc. U.K., 47, 271 (1967).
10. R. T. Emmet, Naval Ship Res. Develop. Center Rep. 2570, 1968.
11. F. A. Richards and R. A. Kletsch in Recent Researches in the Fields of Hydrosphere, Atmosphere, and Nuclear Geochemistry (Sugawara Festival Volume) (Y. Miyake and T. Koyama, eds.), Maruza Co. Ltd., Tokyo, 1964, p. 65.
12. L. Solórzano, Limnol. Oceanog., 14, 799 (1969).
13. J. D. H. Strickland and T. R. Parsons, A Practical Handbook of Seawater Analysis, Fisheries Research Board of Canada, Bulletin No. 167, 1968, p. 84.

13

INORGANIC NITROGEN COMPOUNDS—II. NITRITE

13.1 Introduction

The determination of nitrite in natural waters is conveniently carried out by a diazotization process. Under acid conditions, nitrite ion reacts with an aromatic amine (abbreviated RNH_2) to form a diazo compound, Eqn. (13-1), which is coupled with another aromatic amine ($ArNH_2$) to form an azo dye, Eqn. (13-2). The intensity of the final coloration is directly proportional to the amount of nitrite present:

$$RNH_2 + NO_2^- + 2H^+ \rightarrow \underset{\text{(diazo compound)}}{RNN^+} + 2H_2O \qquad (13\text{-}1)$$

$$RNN^+ + HArNH_2 \rightarrow \underset{\text{(azo dye)}}{RNNArNH_2} + H^+ \qquad (13\text{-}2)$$

Many of the early determinations were made using modifications of the Griess-Ilosvay procedure: Sulfamic acid (RNH_2) and 1-naphthylamine ($HArNH_2$) are used. Most of the reported discrepancies are probably due to variations in quality of reagents and procedures, though there are several interferences: Ferrous iron and other ions (mercuric, silver, lead) may interfere by precipitating, but a 0.5% EDTA solution may be used to complex iron; cupric ion catalyzes the decomposition of the diazo compound; colored ions and turbidity may be removed by flocculation with zinc sulfate and alkali or by using aluminum hydroxide suspension. Some workers feel that the method is rather slow and that the results are not as consistent as might be desired, though optimum conditions have been devised.[1,2]

A more rapid color development and a 5% increase in sensitivity result with the use of sulfanilamide (RNH_2) as the diazotizing agent and N-(1-naphthyl)ethylenediamine dihydrochloride ($HArNH_2$) as the coupling agent, according to Shinn[3] and Bendschneider and Robinson.[4] This procedure is also less sensitive to salinity.

13.2 General Analysis Procedure

Procedure

(a) Reagents

Sulfanilamide. Dissolve 5 g of the compound sulfanilamide, $NH_2SO_2C_6H_4NH_2$, in 500 mℓ of 1.2 N hydrochloric acid (Note 1) and store in an amber, glass-stoppered bottle.

Coupling Reagent, N-(1-Naphthyl)ethylenediamine Dihydrochloride, 0.1% Solution. Purchase the pre-weighed material or prepare as follows: Dissolve 0.50 g of the compound $C_{10}H_7NHCH_2CH_2NH_2 \cdot 2HC\ell$ in distilled water and dilute to 500 mℓ. Store in an amber, glass-stoppered bottle, away from direct sunlight.

Standard Nitrite Solutions. Solution A. Dry pure sodium nitrite at 110° for several hours and cool in a desiccator. Dissolve 0.3450 g of the dried compound in distilled water and dilute to one liter. Add a few drops of chloroform to prevent bacterial growth. Each milliliter contains 5 μg-at or 70 μg of nitrite-nitrogen.

Solution B. Dilute 5 mℓ of Solution A to 500 mℓ with distilled water. Each milliliter of Solution B contains 0.05 μg-at or 0.70 g of nitrite-nitrogen (Note 2).

Solution C. Prepare standard solutions in the range 0.05-1.00 μg-at nitrite-nitrogen/liter as needed by diluting 1-10 mℓ of Solution B to one liter (Note 3).

(b) Analysis of the Sample (Note 4)

To a 50-mℓ sample of Standard Solution C or sea water, add 1.0 mℓ of the sulfanilamide solution. Mix thoroughly.

After two to six minutes, add 1 mℓ of coupling reagent solution. The maximum color intensity develops in about ten minutes and the solution is stable for about two hours.

During this time, measure the absorbance at a wavelength of 543 mμ. Bendschneider and Robinson[4] recommend using 10-cm cells (Note 5), a slit width of 0.10 mm, and distilled water in the reference cell.

Determine the nitrite concentration from a calibration curve prepared by plotting the absorbance versus concentration of Standard Solution C.

NOTES

1. Prepare 1.2 N hydrochloric acid by adding 50 mℓ of reagent-grade concentrated hydrochloric acid (sp. gr. 1.19, 37%) to 250 mℓ of distilled water and diluting to 500 mℓ.

2. Standard Solution A is usually stable for several weeks, but the concentration of Solution B may vary after a few days, and Solution C must be prepared as needed.
3. This solution may be prepared by diluting with distilled water. The absorbance of the final color does not vary significantly with salinity.
4. The temperature should be maintained within 2-3 degrees during the analysis and the preparation of the calibration curve.
5. Also, 2-cm cells may be used.

REFERENCES

1. H. Barnes, Apparatus and Methods of Oceanography, Part 1: Chemical, George Allen and Unwin, London, 1959, Chapter 7.
2. Standard Methods for the Examination of Water, Sewage, and Industrial Wastes, 10th ed., American Public Health Association, Inc., New York, 1955, pp. 246-247.
3. M. B. Shinn, Ind. Eng. Chem. (Anal. Ed.), 13, 33 (1941).
4. K. Bendschneider and R. J. Robinson, J. Mar. Res., 11, 1 (1952).

14

INORGANIC NITROGEN COMPOUNDS—III. NITRATE

14.1 Introduction

Five general approaches have been used to determine nitrate in natural waters.[1,2]

Nitration of Aromatic Organic Compounds. Phenoldisulfonic acid reacts with nitrate ion to produce a nitro derivative, which rearranges in alkaline solution to produce a yellow-colored compound that has a maximum absorption at 410 mμ.[3] There are several interferences: Chloride ion is a negative interference and must be removed by precipitation with silver sulfate, but the precipitation may not be complete and an off color is present when the final color is formed. Nitrite (in concentrations greater than 0.6 mg NO_2^-/liter) is a positive interference. Color and turbidity must be removed by using activated carbon, aluminum hydroxide, or zinc sulfate and alkali. The method is suitable for fresh water, but not sea water, primarily because of the chloride interference. The interference of chloride is minimized in the spectrophotometric determination of nitrate with 2,6-xylenol reagent.[4]

Polarographic Methods. Nitrate ion is catalytically reduced at a dropping mercury electrode in the presence of uranyl ion.[5] There is a linear relation between the diffusion current and the nitrate content above a minimum uranyl ion concentration. Little work on this technique has been reported. The method appears to be unsuitable for shipboard use unless the apparatus is stabilized. Also, there is some question as to the applicability of the method to analysis of natural sea water,[2] although Chow and Robinson[6] examined the method in detail and demonstrated its validity.

Use of Colorimetric Reagents. This approach depends upon the formation of colored products with various reagents which are oxidized by nitrate ion in acid solution. The readily oxidizable reagents include brucine, strychnidine, diphenyl amine, resorcinol, and diphenylbenzidine. Many colorimetric reagents are suitable for determination of nitrate in fresh waters, but lack the sensitivity needed with the lower nitrate concentrations in sea water. Other oxidizing agents, notably nitrate, interfere. Often, the intensity of the final color, which is to be related to the nitrate concentration, is very sensitive to the acid concentration. Finally, the results are often dependent upon trace impurities, e.g., different samples of brucine may result in different intensities with the same nitrate concentration.

Specific-Ion Electrode Determination. Nitrate in fresh water can be determined reliably, as well as rapidly, with a specific-ion electrode.[11] Nitrate concentrations in the range 1-6,000 ppm can be determined directly. Unfortunately, most common anions in sea water interfere with the electrode response, and concentrations of chloride in excess of 500 ppm seriously interfere. Possibly, methods can be devised that will remove interfering anions, though many such procedures involve changes in pH that also adversely affect electrode response.

Reduction of Nitrate. Nitrate ion may be reduced to nitrite, to nitrosyl chloride (NOCℓ), or to ammonia. In some cases, conditions used for reduction to ammonia were too drastic and caused the decomposition of organic nitrogen compounds, but this was corrected by Riley and Sinhaseni.[7] The chief difficulty in reduction to nitrite is finding reagents that will quantitatively reduce nitrate to nitrite and that will not reduce nitrite. A procedure for using hydrazine as a reducing agent has been developed by Mullin and Riley[8] and modified by Strickland and Austin.[9] Henriksen[10] adapted this procedure for use with a Technicon AutoAnalyzer. Reduction with hydrazine has several disadvantages, including the time required for completion (24 hours), difficulty in correcting for large amounts of nitrite, and inhibition by hydrogen sulfide.[2]

It may be concluded by many that no completely satisfactory method for determining nitrate in sea water is presently available. Two reduction procedures are available, however, that seem to be free of most of the difficulties inherent in the methods described above. The procedures consist of the reduction of nitrate to nitrite (copperized cadmium column) or to nitrosyl ion (sulfuric acid). The first procedure is readily adaptable to shipboard use, though preparation and use of the column require some attention to detail. The second procedure is simple and elegant, though it suffers from the disadvantage of use of sulfuric acid and is less sensitive than the first. It requires a smaller sample than the first procedure and is a superior choice for analysis of culture samples and those having higher nitrate concentrations.

14.2 Cadmium-Copper Reduction to Nitrite

In this procedure,[12] a sample treated with ethylenediaminetetraacetate ion (Y^{4-}) is passed through a column of copperized cadmium. Nitrate is reproducibly and nearly quantitatively reduced to nitrite:

$$NO_3^- + Cd + Y^{4-} + H_2O \rightarrow NO_2^- + CdY^{2-} + 2OH^- \quad (14\text{-}1)$$

which is determined by a diazotization procedure (Section 13.2). Copper served as the cathode in the redox couple, and the complexing agent (Y^{4-}) sequesters cadmium and serves as a column conditioner.

Two modifications are used here. The first is a straight column; many workers prefer addition of a self-leveling tube, which has the advantage of preventing the columns from running dry, but which has a dual disadvantage of fragility and holdup. The second modification is the use of a systolic pump to increase the number of samples that can be processed in an hour (Note 1). The two modifications comprise a semiautomated procedure.

Procedure

(a) Reagents

Complexing Solution. Dissolve 38 g of tetrasodium ethylenediaminotetraacetate in distilled water and dilute to one liter.

Column Wash Solution. Dissolve 20 mℓ of complexing solution and 0.125 mℓ of conc HCℓ in distilled water and dilute to one liter.

Copper Sulfate Solution (0.08 M). Dissolve 20 g of $CuSO_4 \cdot 5H_2O$ in 500 mℓ of distilled water.

Hydrochloric Acid (2 M). Pour 85 mℓ of conc. (12 M) acid into distilled water and dilute to 500 mℓ.

Nitric Acid (3 M). Pour 5 mℓ of conc. (12 N) acid into distilled water and dilute to 250 mℓ.

Nitrate Standard Solution. Dissolve 0.5056 g of dried (one hour at 110°) KNO_3 in distilled water and dilute to one liter. Two drops of chloroform may be added as a preservative. The solution should be stored in a dark, tightly stoppered flask. (1 mℓ = 5 μg-at = 70 μg.)

(b) Apparatus

The column (Figure 14-1) now consists of three parts: (1) a sample holder made from 30-mm i.d. borosilicate tubing and tapered to 8 mm i.d.; (2) a feed line, which consists of Tygon tubing connecting the sample holder to a three-way stopcock, the other two ends being connected to waste exit and the reduction column; and (3) the copperized cadmium column. The

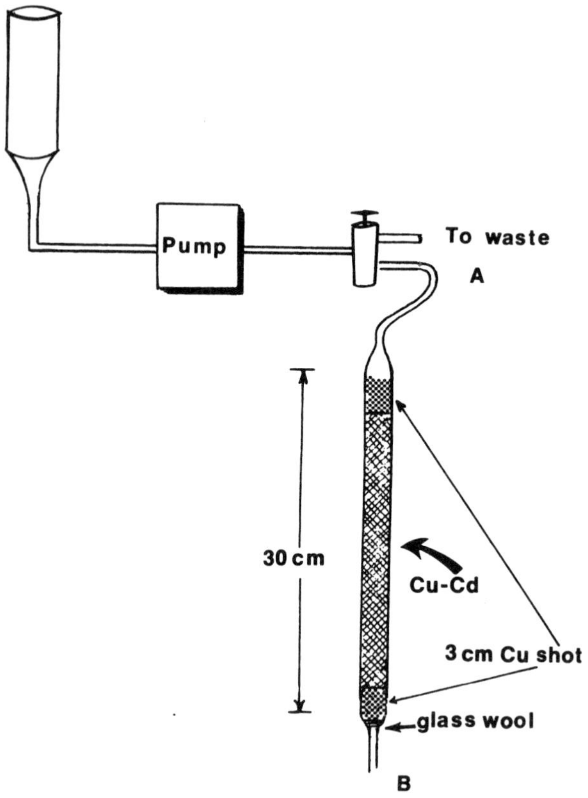

FIG. 14-1

Schematic representation of semiautomated apparatus for nitrate reduction.

feedline of each column is passed through a systolic pump.

(c) Column Preparation

Prepare the column as indicated in Figure 14-1 with a glass wool plug in the bottom, but with the column attached to the reservoir at point A. At this stage, pinch clamps should be available at the top and bottom portions of plastic tubing attached to the column. Fill the column and reservoir with distilled water and remove all air bubbles. Place a rubber stopper attached to a string in the bottom of the reservoir. Then, drain off the water in the reservoir.

Prepare the copperized cadmium column as follows. Place enough washed filings (Note 2) in the reservoir (about 40 g). Add about 100 mℓ of copper sulfate solution to the reservoir and stir the filings until they have a uniform dull gray color, then pull out the stopper by means of the attached string. Gently tap the column to shift the copperized cadmium filings into the column, then clamp at point A. Add copper shot (2 cm), leaving a water void (about 2 cm). Remove the copper sulfate solution and replace with water. With clamps closed at points A and B, assemble the column as shown in Figure 14-1. (DO NOT ALLOW THE FILINGS TO COME IN CONTACT WITH AIR.) Finally, condition the column (Note 3).

(d) Analysis

Rinse the walls of the sample holder with three 10-mℓ portions of column wash solution, with the stopcock to the waste exit and with the pump operating, and drain the sample holder. Stop the pump, and close the stopcock. Add sample (100 mℓ) to the sample holder, followed by 2 mℓ of complexing solution, and mix with a glass rod. Pump the mixture rapidly through the exit line until bubbles are eliminated from the feedline. Then turn the stopcock to the column. Pass 30 mℓ of sample through the column and discard the effluent. Then, collect a 3-5 mℓ sample in a 50-mℓ glass-stoppered mixing cylinder after each discard and shake dry. Then, collect a 20-mℓ sample. The remainder of the sample is used for replicate determinations and any excess is discarded through the exit line. The column must always be kept free of air bubbles and covered with water.

The reduced sample is analyzed for nitrite ion (Section 13.2). Add 1 mℓ of sulfanilamide solution and shake. Wait five to eight minutes, then add 1 mℓ of coupling reagent, and shake. Determine the absorbance at 543 mμ, using a 1-cm cell.

(e) Calibration

The columns should be checked for reduction efficiency E which should be 99 ± 1%:

$$E = A_{NO_3^-}/A_{NO_2^-} \qquad (14\text{-}2)$$

Here, $A_{NO_3^-}$ and $A_{NO_2^-}$ are the absorbances (corrected for reagent blanks) of nitrate and nitrite standards of the same concentration. The reduction efficiency can also be used to correct for nitrite ion that may be reduced by the column.

The columns may be calibrated by using three artificial seawater solutions (diluted to 250 mℓ): (1) 1.00 mℓ of nitrate standard solution; (2) 1.00 mℓ of distilled water; and (3) 1.00 mℓ of nitrite standard solution. Treat 100-mℓ samples from each flask with complexing solution and carry through the analysis procedure, and record the three absorbance values.

Nitrate ion concentration is calculated using the three absorbance values. First, correct the absorbance values for the reagent blank obtained from the absorbance of solution (2). (The value should be less than 0.010 for a 1-cm cell.) Calculate the value of E, Eqn. (14-2), which is the ratio of the corrected absorbances of solutions (1) and (3). Next, calculate the concentration factor F,

$$F = 20/(A_1 - A_2) \tag{14-3}$$

Here, A_1 and A_2 refer to the corrected absorbances of the first and second solutions. The relationship is valid because the absorbance-concentration relationship is linear. Finally, calculate the nitrate concentration $C_{NO_3^-}$ of the sample from the relationship

$$C_{NO_3^-} = FA_3 - 0.99C_{NO_2^-} \tag{14-4}$$

14.3 Reduction to Nitrosyl Chloride[13]

Armstrong's method consists in adding an equal volume of sulfuric acid to a solution which contains both nitrate and chloride and measuring the development of the strong-intensity band at 230 mμ. At this wavelength, Beer's Law is obeyed up to 3 mg NO_3^--N/liter, and at this concentration, the standard deviation is ± 50 μg NO_3^--N/liter. Any organic compounds that absorb at 230 mμ will interfere, but correction for this is made by measuring the absorbance of a treated sample in which nitrosyl chloride has been reduced with hydrazine. Reducing agents interfere, though some organic material is tolerable. Nitrite reacts as nitrate, Eqn. (14-6), and must be corrected for, if present:

$$HNO_3 + 3HC\ell \rightarrow NOC\ell + C\ell_2 + 2H_2O \quad (14\text{-}5)$$

$$HNO_2 + HC\ell \rightarrow NOC\ell + H_2O \quad (14\text{-}6)$$

The method is applicable to fresh water, but chloride must be present in excess of 0.025 M or the final color development will not occur.

Procedure

(a) Reagents

Concentrated Sulfuric Acid. Reagent grade, (98%, sp. gr. 1.84).

Hydrazine Sulfate. Dissolve 2 g of $N_2H_4 \cdot H_2SO_4$ in distilled water and dilute to 100 mℓ.

Standard Nitrate Solution. Solution A (14 μg N/mℓ). Dissolve 0.101 g of KNO_3 in distilled water and dilute to one liter.

Solution B (1.4 μg N/mℓ). Pipet 100 m of Solution A into a one-liter volumetric flask and dilute to the mark with distilled water.

Concentrated Hydrochloric Acid (12 M). Use 37% acid (sp. gr. 1.19).

Dilute Hydrochloric Acid (0.1 M). Dissolve 8.3 mℓ of 12 M hydrochloric acid in distilled water and dilute to one liter.

(b) The Blank Run

For the sample, use 100 mℓ of distilled water to which is added 0.5 mℓ of concentrated hydrochloric acid. The following steps are involved.

In each of two 150 x 20 mm stoppered test tubes, pipet 10.0 mℓ of sample. To one portion, add 0.1 mℓ of hydrazine sulfate solution.

Add 10 mℓ of concentrated sulfuric acid from the dispenser to each test tube (Note 4). Run the sulfuric acid down the side of the test tube. Be sure that the mixing does not cause the solution to boil.

Stopper the test tubes, cool in running water, mix, and cool (Note 5).

Carefully pour the contents of the tube into 5-cm cells; measure the absorbance of each at 230 mμ (Note 6).

If the reagent solutions are essentially free of organic compounds and nitrate ion, the absorbance for a 1-cm cell should be less than 0.1. If this is not the case, the impure reagent must be detected and replaced. Repurify the distilled water. If this is not the source of impurity, purify the sulfuric acid. Heat reagent-grade sulfuric acid to boiling, but do not boil, for 15 minutes, then cool and store in a glass-stoppered bottle.

(c) Analysis of the Sample

The sample may be fresh water or sea water. In either case, the sample should not contain more than 2.5 ppm nitrate-nitrogen, but it must contain more than 2 g of chloride per liter. Dilute samples with high nitrate concentration with 0.1 *M* hydrochloric acid. With freshwater samples, add 0.5 mℓ of 12 *M* hydrochloric acid to 100 mℓ of sample.

The absorbance of the sample and calibration standards should be measured at nearly the same temperature, within two degrees.

Repeat the procedure used in the blank run.

(d) Preparation of the Calibration Plot

The choice of standard concentrations will depend upon the sample. The concentration of nitrate-nitrogen usually is 0-50 μg/liter in sea water and in fresh waters is 0-50 mg/liter depending upon the extent of pollution.

For freshwater samples, dissolve x milliliters of standard nitrate Solution B in artificial sea water (Note 7) and dilute to 100 mℓ. Each sample contains 14x μg N/liter. The following values of x are suggested: 0.0, 0.1, 0.3, 0.5, 0.7, 1.0 (1.5, 2.0, 3.5).

Follow the analysis procedure given, and plot absorbance as a function of concentration. The concentration of an unknown sample may then be determined from the calibration plot. Remember that this procedure gives the total concentration of nitrate and nitrite; if nitrite is present, the concentration must be determined separately and subtracted from the total concentration.

NOTES

1. The modification was suggested by David K. Millard.
2. Prepare washed filings in the following way. Prepare cadmium filings from stick cadmium by using a file and

screening the filings. About 40 g are needed of a fraction that passes through a 2-mm screen and is retained on a 0.5-mm screen. Iron contamination can be eliminated by using a magnet. Wash the cadmium filings in a separatory funnel with hydrochloric acid (2 M), draining and rinsing thoroughly with distilled water. Then, wash with nitric acid (0.3 M) (caution), and rinse with distilled water. Remove nitrate ion by washing with hydrochloric acid (2 M), then rinse thoroughly with distilled water.

3. First, wash with about 50 mℓ of column wash solution. The column should be allowed to stand for about 24 hours and the wash solution should be replaced about three times during this period. Then, pass 3-4 liters of column conditioning solution through the column. This should minimize erratic results and should favor reproducible reduction.
4. The dispenser stopcock should be lubricated only with sulfuric acid or a Teflon stopcock should be used.
5. The test tubes must be stoppered because the absorbant species is fairly volatile.
6. The solution, 50 volume per cent sulfuric acid, is viscous and highly refractive. Take care in the transfer to the cuvette to avoid "schlieren" because of poor mixing. It may be necessary to allow time for small bubbles to clear. The solutions, if stoppered, will be

stable for 24 hours. Light seems to have no effect.

7. Directions for the preparation of artificial sea water are given in Section 4.2(a).

REFERENCES

1. R. S. Lambert and R. J. DuBois, Anal. Chem., 43, 955 (1971).

2. J. P. Riley, in Chemical Oceanography (J. P. Riley and G. Skirrow, eds.), Vol. 2, Academic Press, New York, 1965, Chapter 21.

3. Standard Methods for the Examination of Water, 11th ed., American Public Health Association, Inc., New York, 1955, pp. 175-178.

4. A. M. Hartley and R. I. Asai, Anal. Chem., 35, 1207 (1963).

5. I. M. Kolthoff, W. E. Harris, and G. Matsuyama, J. Am. Chem. Soc., 66, 1782 (1944).

6. D. T. W. Chow and R. J. Robinson, J. Mar. Res., 12, 1 (1953).

7. J. P. Riley and P. Sinhaseni, J. Mar. Biol. Assoc. U.K., 36, 161 (1957).

8. J. B. Mullin and J. P. Riley, Analyt. Chim. Acta, 12, 464 (1955).

9. J. D. H. Strickland and K. H. Austin, J. Cons. Int. Explor. Mer, 24, 446 (1959).

10. A. Henriksen, Analyst, 90, 65 (1965).

11. D. Langmuir and R. L. Jacobson, Environ. Sci. Tech., 4, 834 (1970).

12. E. D. Wood, F. A. J. Armstrong, and F. A. Richards, J. Mar. Biol. Assoc. U.K., 47, 23 (1967).

13. F. A. J. Armstrong, Anal. Chem., 35, 1292 (1963).

15

ORGANIC NITROGEN

15.1 Introduction

In general, total nitrogen (Note 1) in dissolved and particulate matter (retained by a 0.5-μ filter) has been measured by a micro-Kjeldahl procedure (Note 2). The problems and results have been considered in detail by others[1-8] and may be summarized as follows.

Problems Due to the Method. The classical Kjeldahl procedure consists of three steps: (1) organic nitrogen is converted to an ammonium salt by digestion with concentrated sulfuric acid; (2) ammonium salt is converted to ammonia with concentrated sodium hydroxide, and is distilled and (hopefully) recovered quantitatively; and (3) the ammonia in the distillate is determined (typically by titration). There are obvious difficulties of contamination (ammonium compounds in pure sulfuric acid, body nitrogen) or failure to achieve quantitative recovery of ammonia, but most of the difficulties center around the digestion process. As Kirk noted,[4] there are three major problems: (1) sulfuric acid is not a strong oxidizing agent, and oxidation of organic material requires an extended period; (2) pyrolytic decomposition or charring, leading to loss of nitrogen, occurs; and (3) conditions are generally favorable to reduction.

Problems Due to the Sample. There are two major difficulties here. The concentration of dissolved organic nitrogen is very low in sea water (less than 0.1-0.3 mg N/liter) and the possibility of contamination (from the atmosphere, the body, or reagents) is very great. The second difficulty is that many compounds are not readily converted to ammonia during the Kjeldahl procedure. The first difficulty can be overcome to some extent with care and careful assessment of blanks. The second difficulty represents a systematic error and the nitrogen content will be underestimated.

15.2 Micro-Kjeldahl Method

It should be noted that the total nitrogen content may not give the desired information. Many workers wish more specific information (protein content, amino acid composition, and chitin). These determinations have been described by Corner and Cowey.[2] The total nitrogen content gives a measure of the protein content in plankton samples. When there is little non-protein nitrogen present, the protein content is approximately equal to the nitrogen content times 6.25. Possibly a better measure of proteinaceous nitrogen is the determination of so-called albuminoid nitrogen, which measures the ammonia released when alkaline permanganate is used to oxidize the organic material.

The procedure used here is based on the procedure of Miller and Miller,[9] which uses hydrogen peroxide and obviates the need of a catalyst. Small amounts of hydrogen peroxide are added to speed the digestion in a Kjeldahl micromethod. It appears that very rarely are low results obtained, but some hydrogen peroxide contains nitrogen-containing preservative, and high results will be obtained unless blanks are run. Thus the major limitation to the sensitivity of this or any method seems to be the level of nitrogen in the reagents. The procedure here uses the pyrazolone method (Chapter 12) which has a detection limit of about 3 μg NH_3-N/liter.

Procedure

(a) Reagents (Note 3)

Ammonia-Free Water. See Section 12.2.

Sulfuric Acid. Carefully dissolve 100 mℓ of "special" (Note 4) sulfuric acid (96% by weight in 100 mℓ of ammonia-free water. Store in a tightly stoppered glass bottle to prevent atmospheric contamination.

Hydrogen Peroxide. Use reagent-grade 30%.

Antibumping Granules. Fisher's "Boileasers" or Hengar granules must be pretreated (Note 5) if they are used, but certain micro-Kjeldahl units have an electric digester that minimizes bumping.

(b) Analysis (Note 6)

Carry out parallel runs with unknown and standards during the following steps. All runs should be carried out in triplicate.

1. To a micro-Kjeldahl flask, add an accurately known amount of unknown sufficient to produce 2-4 μg of NH_3. If the

sample is likely to foam, add one drop of caprylic alcohol to the unknown and standard solutions (further drops may be added as needed).

2. Add 0.4 mℓ of sulfuric acid solution. Place the flask on the digester and evaporate the solution until white fumes appear. Carry out the digestion (after evaporation of water) for five minutes, then take the bulb of the flask off the heater and allow to cool for about 30 seconds.

3. Hold the flask nearly horizontally and cautiously add two drops or 0.1 mℓ of hydrogen peroxide to the neck of the flask.

4. Return the flask to the heating element and continue the digestion for two minutes. Repeat steps 2-4 (Note 7).

5. Remove the flask from the heating unit and allow the flask to cool to room temperature. Add 10 mℓ of ammonia-free water and mix thoroughly; dilute to 100 mℓ. Continue the analysis by following the pyrazolone method (Chapter 12).

(c) Calculation

Determine the ammonia concentration in the 100-mℓ sample as x μg-at NH_3^--N/liter.

Calculate the organic nitrogen in the original sample as

$$\mu\text{g-at organic N/liter} = \frac{100x\ \mu\text{g-at}}{\text{m}\ell\ \text{sample}} \qquad (15\text{-}1)$$

or calculate the per cent of nitrogen

$$\%\text{N} = \frac{1.4x\ \mu\text{g N} \cdot 100}{\mu\text{g sample}} \qquad (15\text{-}2)$$

15.3 Photochemical Combustion

A photochemical reactor, using a mercury arc lamp, for

oxidation of organic matter, can be used to convert organic nitrogen to nitrite plus nitrate.[10] Samples are analyzed for oxidized nitrogen (nitrite plus nitrate) before and after irradiation. In addition, correction must be made for ammonia that is initially present and subject to oxidation. Detailed analysis of dissolved organic nitrogen in fresh water indicates that careful work requires analysis of ammonia before _and after_ irradiation.[11]

Procedure

(a) Apparatus

See Section 12.3.

(b) Reagents

See Section 12.3.

(c) Analysis

Two 100-mℓ aliquots are required from each sample of filtered sea water. One aliquot is stored in the dark and saved for analysis of initial ammonia and oxidized nitrogen. The second aliquot is treated with two drops of hydrogen peroxide and irradiated. The unit described (Section 11.3) requires 12 hours, though this includes time for oxidation of ammonia nitrogen. A shorter oxidation time (three hours) is feasible for a 1200-W unit. The irradiated sample is cooled, and distilled water added to allow for evaporation.

Following irradiation, the irradiated and unirradiated samples are analyzed for ammonia (Section 12.2) and for oxidized nitrogen (Section 14.2). In the latter analysis, the same column should be used.

(d) Calculations

Calculate the organic nitrogen using the relation

$$\mu\text{g-at N/liter} = (ON_f - ON_i) + (A_f - A_i) \qquad (15\text{-}1)$$

Here, ON_f and ON_i are the oxidized nitrogen concentrations in the irradiated and unirradiated samples, respectively; A_f and A_i have similar designations for the ammonia samples.

NOTES

1. Total nitrogen includes ammonia and organic nitrogen but does not include nitrite and nitrate nitrogen. Organic nitrogen can be determined (a) by volatilizing the ammonia and analyzing the residue or (b) by determining the ammonia and subtracting this value from the total nitrogen value. See Section 36.3 for protein assay.
2. The Conway microdiffusion technique is one example.[6]
3. Other necessary reagents are listed in Chapter 12.2.
4. This should contain less than 2 ppm N and may be purchased (e.g., G. F. Smith Chemical Co.).
5. Strickland and Parsons[3] recommend the following: heat the granules in near-boiling sulfuric acid for 2-3 hours, cool, remove, boil with three changes of water, and dry in the oven at 100° overnight.
6. Review sampling and storage problems in Chapter 1. All glassware should be washed in hot chromic acid-sulfuric acid mixture and rinsed thoroughly with ammonia-free

water before use.

7. For most samples, this should be sufficient; for certain samples, notably vitamins, steps 2-4 must be repeated four times.

REFERENCES

1. J. P. Riley, in Chemical Oceanography (J. P. Riley and G. Skirrow, eds.), Vol. 2, Academic Press, New York, 1965, pp. 398-399.

2. E. D. S. Corner and C. B. Cowey, Oceanogr. Mar. Biol. Ann. Rev., 2, 147-167, (1964).

3. J. D. H. Strickland and R. R. Parsons, A Manual of Sea Water Analysis, Fisheries Research Board of Canada, Bulletin No. 125, (2nd ed.), 1965, pp. 89-93.

4. P. L. Kirk, Anal. Chem., 22, 354 (1950).

5. H. Barnes, Apparatus and Methods of Oceanography, Part I: Chemical, George Allen and Unwin, London, 1959.

6. P. B. Hawk, B. L. Oser, and W. H. Summerson, Practical Physiological Chemistry, 13th ed., McGraw-Hill, New York, pp. 886-888, 1954.

7. C. L. Ogg, in Treatise on Analytical Chemistry (I. M. Kolthoff and P. J. Elving, eds.), Part II, Vol. 11, Interscience, 1965, pp. 461-498.

8. W. W. Umbreit, R. H. Burris, and J. F. Stauffer, Manometric Techniques, 3rd ed., Burgess, Minneapolis, 1959.

9. F. L. Miller and E. E. Miller, Anal. Chem., 20, 481 (1948).

10. F. A. J. Armstrong and S. Tibbits, J. Mar. Biol. Assoc. U.K., 48, 143 (1968).

11. B. A. Manny, M. C. Miller, and R. G. Wetzel, Limnol. Oceanog., 16, 71 (1971).

16

UREA

16.1 Introduction

Urea is an important compound in the nitrogen cycle, though many features of its distribution are uncertain. It is produced by heterotrophic microorganisms and animals, and is consumed chiefly by photosynthetic organisms and, to a degree, by bacterial action. Urea is the chief component of nitrogen excretion by mammals, and a much smaller fraction of nitrogen excretion by zooplankton (about 10% for *Calanus helgolandicus*).[1] Urea accumulation is a balance between phytoplankton uptake and zooplankton activity and excretion. Urea should accumulate during heavy zooplankton grazing, and should be present in low or undetectable concentration during slight or absent zooplankton grazing.

Despite its significance comparatively little is known about urea concentrations in the marine ecosystem. The lack of information is probably a reflection of the fact that comparatively few methods are available for convenient determination of urea in sea water. Newell, Morgan, and Cundy[2] applied the monoxime method that involves complexing urea and diacetyl-monoxime.[2] Emmet[3] described a colorimetric determination using hypochlorite and phenol. McCarthy[4] published a urease method that is simple and can be both specific and sensitive.

16.2 Urease Method for Urea

This method consists in the addition of urease to sea water, warming in water bath at 50° (optimum temperature for rapid enzymatic hydrolysis), and determination of liberated ammonia. The method is specific, but the sensitivity and precision are determined by the ammonia determination, and McCarthy[4] recommended the phenol hypochlorite procedure of Solórzano.[5]

Several features of the determination deserve comment. A buffer is not used, because of interference in the subsequent ammonia determination; the use of an optimum pH probably would not diminish the reaction time (20 minutes) sufficiently to compensate for added difficulties. Composite blanks must be determined for each sample to compensate for traces of ammonia in enzyme preparation, in ammonia-determination reagents, and in the sample. Enzyme inactivation (by boiling) is omitted to avoid turbidity in the ammonia determination, under the conditions of which urease activity is slight. The absorbance-concentration relationship is linear up to 15 μg-at urea N/liter; at concentrations of 1.0 and 3.0 μg-at urea N/liter, the standard deviations were 0.01 and 0.04, respectively.[3]

Procedure

(a) Reagents

The following reagents are prepared using deionized distilled water (Section 8.4).

Complexing Solution. Dissolve 70 g of disodium ethylenediaminetetraacetate in 6.5 liters of distilled water, adjust the pH to 6.5 with 1 M sodium hydroxide solution (Note 1), then dilute to 7 liters.

Clelands Reagent Solution. Dissolve 0.100 g of Clelands Reagent (Note 2) in 65 mℓ of distilled water. Keep frozen.

Urease Solution. Purification of Stock Solution. Use purified (Sigma or Worthington) jack bean urease or purify crude material as follows. Dissolve 0.5 g of material in 22 mℓ of complexing solution and place in a 15-20 cm section of cellulose dialyzer tubing (about 15-mm inflated radius). Suspend the enclosed tube in one liter of complexing solution, refrigerate at 5°, and stir continuously with a Teflon-coated magnetic stirring bar. McCarthy[4] recommends replacing the external complexing solution with another liter every 10-12 hours for 70 hours. After that time, remove the solution in the tube, add 3 mℓ of glycerine, mix, and store at 5°.

Dilute Urease Solution. For analysis, dilute 1 mℓ of purified solution to 100 mℓ with distilled water. Prepare a fresh solution each day.

(b) Sample Preparation

Analyses should be run within an hour of collection. If this is not possible, remove microorganisms by filtration. Store the filtered samples in glass bottles and freeze solid. Under these conditions, the urea concentration should not change for at least three weeks.

(c) Analysis

The enzymatic hydrolysis is effected in the following way. Add 5 mℓ of dilute urease solution to a 50-mℓ sample in a 100-mℓ glass-stoppered mixing cylinder. Place the cylinders in a 50° bath for 20 minutes. Cool the sample to room temperature. Determine the amount of liberated ammonia as soon as possible, to avoid the possibility of atmospheric contamination. Add sequentially to the cooled cylinders phenol solution (2 mℓ), sodium nitroprusside solution (2 mℓ), and oxidizing reagent (5 mℓ), and mix after each. Allow one hour for maximum development and determine (within 4-5 hours) the absorbance at 640 mμ using 10-cm cells.

(d) Calibration

Samples and blanks should be determined in duplicate; a standard run should be performed with each set of analyses. The blank consists of 50 mℓ of sample run in parallel with the test run; but the 5 mℓ of dilute urease solution is added just before the ammonia analysis. The standard run consists of 50 mℓ of standard plus 1 mℓ of urea standard. Calculate the concentration of urea nitrogen in the unknown sample from the absorbance of the unknown corrected for a blank:

$$\mu\text{g-at urea-N/liter} = A_u \times F \qquad (16\text{-}1)$$

The value of the factor F is about 6.5:

$$F = C_s/(A_s - A_b) \qquad (16\text{-}2)$$

C_s is the concentration of the standard, A_s is the absorbance of the standard, and A_b is the absorbance of the blank.

NOTES

1. Dissolve 4 g of NaOH in distilled water and dilute to 100 mℓ.

2. See Sigma Chemical Company, P.O. Box 14508, St. Louis, Mo., 63118.

REFERENCES

1. E. D. S. Corner and B. S. Newell, J. Mar. Biol. Assoc. U.K., 47, 113 (1967).

2. B. S. Newell, B. Morgan, and J. Cundy, J. Mar. Res., 25, 201 (1967).

3. R. T. Emmet, Anal. Chem., 41, 1648 (1969).

4. J. J. McCarthy, Limnol. Oceanog., 15, 309 (1970).

5. L. Solórzano, Limnol. Oceanog., 14, 799 (1969).

17

COPPER

17.1 Introduction

In natural waters, copper occurs principally as soluble ionic species, though a small fraction of particulate copper is also present.[1] Williams[2] and others[1,3] have provided evidence that copper (commonly 5-28%, though as much as 50% of the total) is associated with organic material in sea water.[1] The nature of the species is a matter of speculation.[1-3]

Recent research has been concerned with the direct analysis of trace amounts of copper by atomic absorption spectroscopy. For example, Ramakrishna and co-workers[4] and Fabricand and co-workers[5] reportedly were able to analyze for 1-2 ppb of copper. Magee and Rahman[6] were able to analyze for 2.5 ppb of copper in sea water after extracting a copper complex of ammonium pyrrolidine dithiocarbamate (APDC) into ethyl acetate. Brooks and co-workers[7] determined "soluble" copper by extraction of complexes of APDC into methyl isobutyl ketone, with a limit of detection of about 0.10 ppb.

As atomic absorption accessory equipment is not available in many laboratories, the method of choice is probably one of the spectrophotometric procedures. All of these procedures depend upon formation of a colored complex of copper, extraction of the copper complex into a suitable organic solvent, and spectrophotometric determination of concentration. Diquinolyl is suitably sensitive and is more specific than dithizone and is probably less capricious.[8] Neocuproine (2,9-dimethyl-1,10-phenanthroline) was found to be more soluble than diquinolyl, and somewhat more sensitive and more specific for copper.[1] With neocuproine, it is possible to determine both total soluble and particulate copper.[1] The distinction between the two forms of copper is, of necessity, an arbitrary one.

Neocuproine (Section 17.3) will probably replace sodium diethyldithiocarbamate, which has been a standard complexing agent for determining the concentration of copper in natural waters.[9] A modification of a procedure developed by Chow and Thompson[9] is used here. This procedure consists in forming the 1:2 copper-diethyldithiocarbamate complex, extracting the complex into xylene, and measuring the absorbance of the xylene solution. The useful lower limit of this procedure is reported[9] to be 0.002 μg-at Cu/liter or about 0.1 ppb.

17.2 Diethyldithiocarbamate Method[9]

Procedure

(a) Reagents

Distilled Water. Copper-free distilled water is obtained by redistilling distilled or deionized water from an all-glass still (see Section 8.4).

Xylene. Use analytical grade.

Sodium Diethyldithiocarbamate (SDDC) 1% (w/v). Dissolve 1.0 g of reagent-grade SDDC in copper-free distilled water and dilute to 100 mℓ. Chow and Thompson[9] recommend storing the reagent at a pH of 10. The reagent is moderately stable but should be renewed every few weeks to obtain maximum color development and avoid high reagent blanks.

Artificial Sea Water. This is prepared as described in Chapter 4.

Standard Copper Solution. Solution A (125 μg Cu/mℓ. Add 0.1250 g of fine granular copper metal (reagent grade) to 3 mℓ of concentrated nitric acid (69%, sp. gr. 1.42) and cautiously add 3 mℓ of concentrated sulfuric acid (96%, sp. gr. 1.84). Heat until dense white fumes are produced. Allow the solution to cool. Dilute to one liter with redistilled water.

Solution B (0.125 μg Cu/mℓ). Dilute 10.0 mℓ of Solution A to one liter with distilled water.

Solution C (0.125 μg Cu/mℓ). Dilute 10.0 mℓ of Solution B to 100 mℓ with distilled water.

(b) Collection of Samples

Use nonmetallic samplers of the Niskin or Van Dorn type suspended from a stainless steel wire to obviate contamination.[1] Surface water samples may be collected by immersing a well-seasoned Pyrex bottle below the surface by means of a weighted rope. Transfer the sample immediately to polyethylene storage bottles.

(c) Pretreatment of the Sample

If the seawater sample contains much plankton, an emulsion will form with the xylene during the analysis. Remove the plankton from a known volume of water by filtration through a sintered glass filter or a membrane filter before continuing with the analysis.

(d) Analysis of the Sample

Measure 500 mℓ of sample into a polyethylene bottle (Note 1). Add 2.0 mℓ of 1% SDDC reagent. Add 5.0 mℓ of xylene. Extract the copper complex by shaking the bottle for one hour using a Burrell shaker. Pipet out the (upper) layer with a medicine dropper and transfer to a 1-cm cell. Measure the adsorption of the xylene extract at 436 mμ, using a 0.05 mm slit width. Use pure xylene in the reference cell (Note 2). Obtain the copper concentration by referring to the calibration curve.

(e) Reagent Blank

A reagent blank should be checked periodically by using 500 mℓ of artificial sea water and following the analysis procedure.

(f) Calibration Curve

Add a known amount of standard copper Solution B (Note 3) to a polyethylene bottle and dilute to 500 mℓ with artificial sea water. Use the analysis procedure given above and record the observed absorbance. Plot the observed absorbance versus added concentration of copper (μg-at Cu/liter). The slope of the linear variation of absorbance with concentration will be constant from one sample to another and should be about 3.35 absorbance units μg-at Cu/liter. The blank will vary with the particular seawater sample used and a corresponding correction should be made when reading the copper concentration directly from the calibration curve.

17.3 Neocuproine Method[1]

Neocuproine, Ch, forms an orange-colored 1:2 compound with cuprous copper, $CuCh_2^+$ in a pH range of 3-10.[1,10] Few metals and anions interfere. Chromium interferes, though in the absence of iron, and the interference of beryllium can be obviated by the proper addition of reagents.[1]

Particulate copper analysis has some special problems. This form is present in slight amounts, and a large volume of water must be filtered. Contamination by several types of filters has been reported,[10] though glass filter pads may eliminate the contamination by the filters. Digestion with dilute hydrochloric acid is evidently incomplete, and perchloric acid is a more effective digestion reagent.[1]

In this procedure, a sample containing copper is treated with a reducing agent to produce cuprous ion in the presence of neocuproine, the pH of the mixture is adjusted to 5.5 with buffer, and the concentration is determined colorimetrically. The method is applicable to freshwater samples since no salt error is observed.

Procedure

(a) Reagents (Note 4)

Neocuproine Solution. Dissolve 1.0 g of neocuproine (Note 5) in one liter of redistilled ethanol.

Hydroxylamine Hydrochloride. Dissolve 100 g of

hydroxylamine hydrochloride in 600 mℓ of redistilled water and filter through a Whatman GF/C filter pad. Add 5 mℓ of neocuproine solution and extract with 25 mℓ of chloroform. Continue the extractions until the chloroform layer remains colorless, then extract with 25 mℓ of carbon tetrachloride. Dilute the aqueous layer to one liter with redistilled water.

Sodium Acetate Solution. Dissolve 453 g of sodium acetate in 800 mℓ of distilled water. Filter through a Whatman GF/C filter pad. Add 2 mℓ of hydroxylamine hydrochloride solution and 5 mℓ of neocuproine solution.

Extract the solution with 25 mℓ of chloroform and continue the extractions until the chloroform layer remains colorless. Make a final extraction with 25 mℓ of carbon tetrachloride. Dilute the aqueous layer to one liter with distilled water.

Digesting Reagent. Use 70-72% perchloric acid (caution).

(b) Analysis of Particulate Copper

At least two liters of sea water, collected in a nonmetallic sampler, is required. Filter the sample through a membrane filter (e.g., Millipore Type HA) having a pore size of 0.5 μ (Note 6). The filter can be placed in a glassine envelope for subsequent analysis. The filtrate is used for analysis of total copper.

Place the filter in the bottom of a 50-mℓ glass-stoppered mixing cylinder; add 3 mℓ of digesting reagent. Heat the cylinder in hot water or under an infrared lamp until the solution is clear. Allow the solution to cool and continue with the analysis.

Add 5 mℓ of sodium acetate solution followed by 2 mℓ of hydroxylamine hydrochloride solution and 5 mℓ of neocuproine. Mix well. Dilute to 25 mℓ with distilled water. The pH should be approximately 5.5. Transfer the treated solution to a spectrophotometer cell and measure the absorbance at 454 mμ (Note 7). Prepare a blank solution from the reagents used in the analysis procedure and place the blank solution in the reference cell.

(c) Analysis of Total Soluble Copper

Place 25 mℓ of filtered sea water in a 50-mℓ glass-stoppered mixing cylinder. Add 3 mℓ of digesting reagent and digest as before. Allow the cylinder and contents to cool to room temperature. Add 2 mℓ of hydroxylamine hydrochloride solution and 5 mℓ of neocuproine solution. Mix well. Dilute to 50 mℓ with sodium acetate solution. Mix well. The pH should be approximately 5.5. Transfer the contents of the cylinder to a spectrophotometer cell (Note 7) and measure the absorbance at 454 mμ, corrected for a reagent blank as in the preceding section.

(d) Calibration Curve

Prepare calibration solutions as follows: Place in a 50-mℓ mixing cylinder, 2 mℓ of hydroxylamine hydrochloride, 5 mℓ of neocuproine solution, and 5 mℓ of sodium acetate solution. Use a microburet or a 1-mℓ pipet which is graduated in 0.1-mℓ intervals to add x milliliters of standard

Solution B (Note 8). Dilute to 25 mℓ with distilled water for the particulate copper curve; dilute to 50 mℓ for the total soluble copper curve. Measure the absorbance as before, and plot the absorbance as a function of concentration (Note 9).

Determine the copper content of the unknown sample by reference to the calibration curve.

NOTES

1. Polyethylene bottles should be thoroughly washed with xylene because new bottles sometimes give an emulsion with xylene during the extraction.
2. The color intensity will fade in strong light or when an excess of SDDC is not present. Otherwise, the absorbance at 436 mμ should be constant for 48 hours.
3. Suggested amounts: x milliliters of Solution B, where x is 0.0, 1.0, 2.0, 3.0, 5.0, 7.5, 9.0, 10.0. The concentration is 2.50 x μg Cu/liter.
4. See Section 17.2 for directions on standard copper solutions and distilled water.
5. Neocuproine is 2,9-dimethyl-1,10-phenanthroline and may be purchased from G. Frederick Smith Chemical Co., Columbus, Ohio 43222.
6. Other workers[1] have used glass filter pads. A filter pad or a membrane filter should be checked for copper content in a blank run.
7. These solutions follow Beer's Law in the concentration

range 0.15-10 μg Cu/mℓ. A 1- or 2-cm cell may be used in the particulate copper determinations; a 10-cm cell is needed for the total soluble copper determinations.

8. Distilled water may be used for the calibration solutions because there is no significant salt error. The following values of x are suggested: 0.0, 1.0, 2.0, 3.0, 4.0, 5.0, 6.0, 7.0, 8.0.

9. The concentrations are: total soluble copper, 25x μg Cu/liter; particulate copper: 1.25x μg Cu/two liters or 0.625x μg Cu/liter, where x is the milliliters of Solution B. Also, 63.5 μg Cu = 1 μg-at Cu.

REFERENCES

1. J. E. Alexander and E. F. Corcoran, Limnol. Oceanog., 12, 236 (1967).
2. P. M. Williams, Limnol. Oceanog., 14, 156 (1969).
3. J. F. Slowey, L. M. Jeffrey, and D. W. Hood, Nature, 214, 377 (1967).
4. T. V. Ramakrishna, J. W. Robinson, and P. W. West, Anal. Chim. Acta, 37, 20 (1967).
5. B. P. Fabricand, R. R. Sawyer, S. G. Ungar, and S. J. Adler, Geochim. Cosmochim. Acta, 26, 1023 (1962).
6. R. J. Magee and A. K. M. Rahman, Talanta, 12, 409 (1965).
7. R. R. Brooks, B. J. Presley, and I. R. Kaplan, Talanta, 14, 809 (1967).
8. J. P. Riley and P. Sinhaseni, Analyst, 83, 299 (1958).
9. T. J. Chow and T. G. Thompson, J. Mar. Res., 11, 124 (1952).
10. K. T. Marvin, R. R. Proctor, and R. A. Neal, Limnol. Oceanog., 14, 320 (1970).

18

MANGANESE

18.1 Introduction

A common method for determination of trace quantities of manganese depends upon the oxidation of this element to permanganate ion and the colorimetric estimation of this intensely colored ion.[1,2] Oxidation is effected with persulfate or periodate, though the latter reagent seems to be more satisfactory.

The direct analysis of manganese in natural waters has been achieved[3] by atomic absorption spectroscopy, though the method is directly applicable only to samples of relatively high manganese content. The nominal sensitivity of a typical atomic absorption spectrophotometer for manganese is 0.04 ppm,[4] though the concentration in sea water is as low as 0.0005 ppm.

A third method of direct estimation is based on the catalytic action of manganous ion on an organic base. Harvey[5] based his procedure on the oxidation of "tetra-base" (_p_,_p_'-tetramethyldiaminodiphenylmethane). The color which is produced fades rapidly at room temperature, and it seems unlikely that this reagent can be used for routine work. Another procedure, using malachite green, has been described by Fernandez and co-workers.[6] More recently, Strickland and Parsons[7] have given the details for the direct estimation of manganese in sea water by "leuco-base" (Note 1) method. A modification of this procedure is given here.

18.2 Leuco-Base Method

The method consists in treating a sample of water which contains 0.0025-0.25 μg-at Mn/liter with acetate buffer, followed by oxidizing agent (periodic acid), followed by leuco-base. The oxidation of the leuco-base to a dye (malachite green) is catalyzed by manganous ion, and the amount of dye present after several hours is related to the amount of manganese present in the original sample. It is necessary to maintain the pH at a constant value (4-4.2), to control the temperature bath, and to adjust the salinity to within 2-3 $^{o}/oo$ for all samples and calibration solutions.

Procedure[7]

(a) Reagents

Manganese-Free Sea Water. Pass a liter of sea water through a membrane filter (e.g., Millipore, 0.5 μ) and pour the water in a clean beaker. Add 5% sodium hydroxide solution (Note 2) dropwise with vigorous stirring to the filtered sea water until a slight amount of precipitate is formed. The precipitate should not redissolve on stirring for five minutes. Heat the solution to the boiling point and boil for five minutes. Cover the beaker and allow to stand at room temperature for 2-3 hours. Decant most of the liquid and filter through filter paper (Whatman #12). Then, refilter through a membrane filter. Acidify the filtrate to a pH of 7.8-8.2 with dilute hydrochloric acid (1 mℓ of concentrated acid in 100 mℓ of distilled water). Dilute the treated sea water with distilled water until a salinity of about 29°/oo is obtained. Store the manganese-free sea water in a tightly sealed polyethylene bottle.

Acetate Buffer. Dissolve 30 mℓ of glacial acetic acid (reagent grade) in 800 mℓ of glass-distilled water. Add 5% sodium hydroxide solution dropwise with shaking until the pH of the buffer is about 4.1-4.2 (Note 3). Dilute to one liter.

Potassium Periodate Solution. Dissolve 0.5 g of analytical reagent KIO_4 in 250 mℓ of distilled water. Add one small pellet (about 0.2 g) of potassium hydroxide. Shake well and store in an amber glass bottle (Note 4).

Leuco-Base Reagent. Dissolve 0.10 g of pure leuco-base (Note 1) in 125 mℓ of pure acetone (Note 5).

Standard Manganese Solution. Solution A (1.50 μg-at Mn/mℓ). Dissolve 0.255 g of manganous sulfate ($MnSO_4 \cdot H_2O$) in redistilled water and dilute to one liter. Add two drops of concentrated hydrochloric acid (12 M). The solution is stable indefinitely.

Alternatively, weigh out 0.298 g of manganous chloride ($MnC\ell_2 \cdot 4H_2O$) in redistilled water and dilute to one liter. Determine the chloride concentration of a 100-mℓ sample by a potentiometric titration with silver nitrate. Calculate the exact manganese concentration. Add two drops of concentrated hydrochloric acid after the determination.

<u>Solution B (0.00150 μg-at Mn/mℓ)</u>. Dilute 1.0 mℓ of Solution A to one liter with redistilled water.

(b) <u>Collection of the Samples</u>

Use nonmetallic samplers and follow the precautions outlined for the collection of samples for copper analysis. If the samples are not to be analyzed within a few hours, freeze the samples. Remove particulate material from the sample with a membrane filter (e.g., Millipore HA, 0.45 μ).

(c) <u>Analysis of Sample</u> (Note 6)

Pour exactly 30 mℓ of sample into a 50-mℓ graduated mixing (unknown) cylinder equipped with a glass stopper. To a second (standard) cylinder, add exactly 30 mℓ of sample and pipet in 1.0 mℓ of manganese solution B. To a third (blank) cylinder, add 30 mℓ of manganese-free sea water. To each cylinder, measure enough acetate buffer to make the total volume in each cylinder 45 mℓ (Note 7).

Place the cylinders in a thermostated bath which is kept at 25°C. Allow the cylinders to stand in the bath for 15-20 minutes.

Add 5.0 mℓ of KIO_4 solution from an automatic pipet to each cylinder. Stopper the cylinder and mix well. Return the cylinder to the bath for 10-15 minutes. Into each cylinder, pipet 1.0 mℓ of leuco-base with rapid mixing. Return to the constant-temperature bath for 4-5 hours from the time of addition of the last reagent.

Measure the absorbance at 615 mμ of a 2-cm cell versus

absorbance of the unknown cylinder at 615 mℓ using the blank solution in the reference cell. Repeat using the standard cylinder.

(d) Calculation

Calculate the concentration in microgram-atoms of manganese per liter from the relationship (Chapter 8, Eqn. 14)

$$\text{g-at Mn/liter} = \frac{A}{A_s - A} \times C_i \qquad (18\text{-}1)$$

where A is the absorbance of the unknown solution corrected versus a "blank" solution and corrected for any cell-to-cell blank; A_s is the absorbance of the "standard" solution; and C_i is the concentration of solution B in the cylinder (0.05).

NOTES

1. Leuco-base, p,p'-benzylidene bis(N,N-dimethylaniline), is oxidized to malachite green.
2. Dissolve 5 g of reagent-grade sodium hydroxide in 95 mℓ of distilled water.
3. Remove a sample of the solution and check the pH with a glass and calomel electrode pair.
4. Best results are obtained when the solution is freshly prepared.
5. The solution is fairly stable; avoid evaporation.
6. Glassware used in the determination should be cleaned by rinsing with a few drops of concentrated hydrochloric acid, then washing thoroughly with distilled water.
7. Duplicate samples should be run for precise work. For a given station, duplicate "standard" samples should be

sufficient. The manganese-free sea water has the appropriate salinity for most work, though for very precise work, the salinity should be within $3^{o}/oo$ of the sample salinity. This is particularly true for low manganese levels.[7]

REFERENCES

1. Standard Methods for the Examination of Water and Wastewater, 11th ed., American Public Health Association, Inc., New York, 1960, pp. 155-170.

2. T. G. Thompson and T. L. Wilson, J. Am. Chem. Soc., 57, 233 (1935).

3. B. P. Fabricand, R. R. Sawyer, S. G. Ungar, and S. J. Adler, Geochim. Cosmochim. Acta, 26, 1023 (1962).

4. FWPCA Methods for Chemical Analysis of Water and Wastes, Analytical Quality Control Laboratory, Cincinnati, Ohio, November 1969, p. 87.

5. H. W. Harvey, J. Mar. Biol. Assoc. U.K., 28, 155 (1949).

6. A. A. Fernandez, C. Sobel, and S. L. Jacobs, Anal. Chem., 35, 1721 (1963).

7. J. D. H. Strickland and T. R. Parsons, A Manual of Sea Water Analysis, Fisheries Research Board of Canada, Bulletin No. 167, 1968.

19

"SOLUBLE" AND "PARTICULATE" IRON

19.1 Introduction

Experimentally, it is possible to differentiate two forms of iron in natural waters: "soluble" and "particulate". Lewis and Goldberg[1] initiated the common practice of designating the "soluble" form as the iron compounds that pass through a 0.5-μ membrane filter. These workers devised an excellent procedure for iron determinations using membrane filters and α,α-dipyridyl for "particulate" iron and bathophenanthroline for "soluble" iron. For both analyses, fuming perchloric acid was used to bring both forms of iron into solution.

Strickland and Austin[2] suggested the practice of subdividing both the "soluble" and "particulate" fractions into biologically reactive and unreactive forms. This was achieved by a treatment that rapidly dissolved all forms of precipitated or colloidal ferric hydroxide, but did not dissolve such materials as unreactive clays and muds, ferric silicate compounds, and ignited iron oxides. Even though this was an arbitrary subgrouping, Strickland and Austin[2] believed that the amount of iron brought into solution by their treatment had a more direct relation to the amount of immediate use to growing plant cells than does the total iron content of a sample.

Collins and Diehl[3] have suggested an additional modification: the use of 2,4,6-tripyridyl-s-triazine (TPTZ) instead of bathophenanthroline. Not only is TPTZ considerably cheaper, but it has a comparable sensitivity. The TPTZ reacts with iron (in the ferrous form) to yield a violet complex, which in the presence of perchlorate ion, may be quantitatively extracted with nitrobenzene as the compound $Fe(TPTZ)_2(C\ell O_4)_2$.

19.2 TPTZ Method

Procedure

(a) Reagents

Deionized Water. See Section 8.4.

TPTZ Solution. Dissolve 0.312 g of the tan, crystalline solid TPTZ (Note 1) in a few drops of concentrated hydrochloric acid and dilute to one liter with deionized water.

Nitrobenzene. Use reagent-grade or specially purified material (Note 2).

Sodium Acetate-Acetic Acid Buffer. Dissolve 82 g of sodium acetate and 58 mℓ of glacial acetic acid in deionized water and dilute to one liter (Note 2). Transfer the solution to a large separatory funnel. Add 10 mℓ of TPTZ solution to a large separatory funnel. Add 10 mℓ of TPTA solution, 10 mℓ of hydroxylammonium chloride-sodium perchlorate solution, and 25 mℓ of nitrobenzene (Note 3). Shake and allow the layers to separate. Withdraw the nitrobenzene layer and discard. Repeat the extraction with a second 25-mℓ portion of nitrobenzene.

Hydroxylammonium Chloride-Sodium Perchlorate Solution. Dissolve 100 g of sodium perchlorate and 100 g of hydroxylammonium chloride (hydroxylamine hydrochloride) in deionized water, and dilute to one liter. Transfer the solution to a large separatory funnel, preferably one with a Teflon stopcock. Add 10 mℓ of TPTZ solution and 25 mℓ of nitrobenzene. Shake and allow the layers to separate. Withdraw and discard the lower nitrobenzene layer. Repeat the extraction with nitrobenzene to make certain all of the iron has been removed.

Standard Iron Solution. Solution A. Accurately weigh a 0.1000-g piece of electrolytic iron (Note 4) and dissolve in 20 mℓ of concentrated hydrochloric acid. Quantitatively, transfer the solution to a one-liter volumetric flask and dilute to one liter. Mix thoroughly.

Solution B (2500 μg Fe/liter). Pipet 25.0 mℓ of Solution A into a one-liter volumetric flask, add 5 mℓ of concentrated hydrochloric acid and dilute to one liter with deionized water.

Solution C (50 μg Fe/liter). Pipet 20.0 mℓ of solution B into a one-liter flask, add 5 mℓ of concentrated hydrochloric acid, and dilute to one liter with deionized water.

(b) Collection of Samples

For the determination of "particulate" iron in sea water, samples of about one liter should be collected. Use nonmetallic samplers suspended from a stainless steel wire to guard against contamination. Burton and Head[4] recommend that immediately after filtration, samples be transferred to polyethylene bottles and acidified (0.5 mℓ of high-purity hydrochloric acid per liter of sample).

(c) Pretreatment of Samples (Note 5)

"Particulate" iron: Pass a measured sample of about one liter through a membrane filter (e.g., Millipore Type HA) having a pore size of 0.5 μ. Place the filter in the bottom of a 50-mℓ glass-stoppered measuring cylinder. Add 10 mℓ of digesting solution (Note 6) and heat the cylinder in a bath of boiling or near-boiling water for 10 minutes. Allow the cylinder to cool and continue the analysis.

"Soluble" iron: Add 10 mℓ of hydrochloric acid digesting reagent to 100 mℓ of filtered sample. Allow the mixture to stand for about 15 minutes, and continue with the analysis.

(d) Analysis of the Sample[3]

Pipet 100 mℓ of the pretreated water into a 125-mℓ separatory funnel. Add 2.0 mℓ of iron-free hydroxylammonium chloride-sodium perchlorate solution, 5.0 mℓ of TPTZ solution, and 5.0 mℓ of iron-free buffer solution (Note 7). Shake the separatory funnel for about two minutes. Add 10 mℓ of nitrobenzene, shake for one minute, and allow the phases to separate. Gently swirl the funnel to dislodge drops of nitrobenzene which cling to the upper walls. Drain the nitrobenzene layer into a volumetric flask. Dilute the nitrobenzene extracts to 25.0 mℓ with ethanol. Determine the absorbance of the solution at 595 mμ, using 5-cm cells. Use a solution of 20 mℓ of nitrobenzene and 5 mℓ of ethanol in the reference cell.

(e) Reagent Blank

Use 100 mℓ of deionized water or 100 mℓ of artificial sea water, and carry through the entire analysis. Subtract the absorbance of the blank from the absorbance of the unknown.

(f) Calibration Curve

Pipet 5.0 mℓ of a standard iron Solution C into the 125-mℓ separatory funnel, dilute the solution to 100 mℓ with deionized water or artificial sea water, and carry out the analysis as described. Repeat using various volumes ranging from 0 to 50. Plot each absorbance as a function of the concentration (μg Fe/liter). A linear relation is obtained, but the line does not pass through the origin.

Knowing the absorbance of the unknown solution, read off the concentration of the unknown sample.

NOTES

1. TPTZ (2,4,6-tripyridyl-s-triazine) has a melting point of 242-243.5° and may be obtained from the G. Frederick Smith Chemical Company, Columbus, Ohio.
2. Suitable nitrobenzene and electrolytic iron may be obtained from the G. Frederick Smith Chemical Company, Columbus, Ohio.
3. Avoid breathing nitrobenzene vapors. Use a well-ventilated area. If any nitrobenzene comes in contact with the skin, wash it off immediately and thoroughly.
4. Pyrex-brand glass or equivalent should be used. It is

helpful to allow the reagents to stand in the bottles that will be used for storage a week before removing traces of iron from the reagent solutions.

5. An alternative digestion procedure consists in treating a 100-mℓ sample with thioglycollic acid (1 mℓ of purified[5] material). The treated material is heated in sealed polyethylene bottles at 80° for two hours.

6. The digesting solution is prepared by adding 20.0 mℓ of concentrated hydrochloric acid (sp. gr. 1.19) to deionized water and diluting to 500 mℓ.

7. The pH should be between 4 and 5 at this point. If not, neutralize with a few drops of ammonium hydroxide.

REFERENCES

1. G. L. Lewis, Jr. and E. D. Goldberg, J. Mar. Res., 13, 183 (1954).

2. J. D. H. Strickland and K. H. Austin, J. Cons. Int. Explor. Mer, 24, 446 (1956).

3. P. Collins and H. Diehl, J. Mar. Res., 18, 152 (1960).

4. J. D. Burton and P. C. Head, Limnol. Oceanog., 15, 164 (1970).

5. J. A. Tetlow and A. L. Wilson, Analyst, 89, 442 (1964).

20

MOLYBDENUM

20.1 Introduction

Molybdenum is thought by many to be an important micronutrient in natural waters. This element is a cofactor for enzymes involved in nitrate reduction. Molybdenum, together with iron, appears to be an important component in enzymes associated with nitrogen fixation. Goldman[1,2] suggested a shortage of molybdenum can limit photosynthesis in lakes. The possibility of the molybdenum-limitation of marine photosynthesis is important, but difficult to verify. In part, the difficulty may be ascribed to the unavailability of a convenient, sensitive, accurate method.

Most procedures depend upon precipitation or coprecipitation of molybdenum, followed by analysis. In general, these procedures tend to be tedious and time-consuming. One improvement consists in complexing the molybdenum with 8-quinolinol (oxine) and coprecipitating with aluminum sulfate at controlled pH using optimum concentrations, and adaptation to routine use.[3] Another improvement involves a rapid absorbing colloidal flotation method for separating molybdenum.[4]

20.2 Modified Oxine Method

Molybdenum in seawater samples is complexed with oxine (8-quinolinol) coprecipitated with aluminum hydroxide, separated, and determined colorimetrically as a thiocyanate complex.[3] The procedure is sensitive enough to measure molybdenum in a useful range, 0.5-10 μg Mo/liter, and no salt effect is noted. The critical factor is control of the pH at which coprecipitation is effected.

Procedure

(a) Reagents

Distilled Water. See Section 8.4.

Acetic Acid (2 N). Dilute 166 mℓ of glacial acetic acid to one liter.

Hydrochloric Acid (1.5 N). Dilute 80 mℓ of concentrated acid (37%) to one liter.

Sulfuric Acid (18 N). Cautiously pour conc. H_2SO_4 (96%) into an equal volume of distilled water.

Sulfuric Acid (1 N). Dilute 55 mℓ of 18 N acid to one liter.

Mixed Acid Solution. Cautiously add 50 mℓ of conc. HNO_3 (69%) to 50 mℓ of 70% $HC\ell O_4$. Store in a plastic wash bottle.

Oxine Solution. Dissolve oxine (5.0 g) in 100 mℓ of 2 N acetic acid. Filter through a sintered glass filter (F). The clear yellow solution should be stable for several months.

Aqueous Ammonia (7.5 M). Dilute 100 mℓ of conc. ammonia (28%) with an equal volume of distilled water.

Aluminum Sulfate Solution. Dissolve 21 g of $A\ell_2(SO_4)_3 \cdot 18H_2O$ in 280 mℓ of distilled water.

Ferric Chloride Solution. Dissolve 0.5 g of $FeC\ell_3 \cdot H_2O$ in

100 mℓ of 1 N H_2SO_4.

Tannic Acid Solution. Dissolve 5 g of tannic acid in 95 mℓ of distilled water.

Extraction Mixture. Mix equal volumes of amyl alcohol and carbon tetrachloride (avoid inhaling and work in a well-ventilated area).

Potassium Thiocyanate Solution. Dissolve 40 g of KSCN in 60 g of water.

Stannous Chloride Solution. Dissolve 40 g of $SnC\ell_2 \cdot H_2O$ in 60 mℓ of 1.5 N HCℓ.

Molybdate Standard. Dissolve 1.0845 g of pure ammonium molybdate, $(NH_4)_6Mo_7O_{24} \cdot 4H_2O$, in 500 mℓ of distilled water (Note 1).

(b) Analysis

Treat 500 mℓ of filtered sample (Note 2) successively with (1) 2.5 mℓ of 18 N H_2SO_4, (2) 2.5 mℓ of $A\ell_2(SO_4)_3$ solution, (3) 2.0 mℓ of $FeC\ell_3$ solution, and (4) 5 mℓ of oxine reagent. Shake well. Adjust the pH to 4.0 with 7.5 M aqueous ammonia, then add tannic acid solution (2 mℓ) and shake well.

Collect and redissolve the precipitate as follows. The precipitate is collected on a Buchner funnel (Whatman No. 42 filter paper). Dissolve the precipitate in 10 mℓ of HNO_3-$HC\ell O_4$ mixture. Evaporate the solution to dryness using a heat lamp or a low-temperature heater. Redissolve the residue in 15 mℓ of distilled water and transfer to a 125-mℓ separatory funnel.

The solution is analyzed using Sandell's thiocyanate procedure.[5] Add 10 mℓ of 6 N HCℓ and dilute to 50 mℓ. Add 2.5 mℓ of extraction mixture and shake vigorously. Separate and discard the lower organic layer. Add successively with shaking KSCN solution (1 mℓ), $SnCℓ_2$ solution (1 mℓ), and extraction mixture (1 mℓ). Shake vigorously for one minute. Remove the lower layer after separation and drain into a 1-cm cell. Record the absorbance at 470 mμ relative to a blank.

(c) Calculations

Prepare a calibration curve by treating uncontaminated one-liter samples with 0, 5, 10, 15, and 20 μg of Mo and correcting the mean absorbance values for blank absorbance. From this plot determine the slope m. Calculate the factor F (= 1/m). Calculate the concentration of an unknown solution from the absorbance of the unknown solution using absorbance A_u corrected for blank absorbance A_b:

$$\text{conc.}(\mu\text{g Mo/liter}) = (A_u - A_b)F \qquad (20\text{-}1)$$

The value of F is about 55.5.

NOTES

1. Alternatively, dissolve 0.3015 g of dry MoO_3 (99.5%) in 5 mℓ of 1 M NaOH (40 g NaOH/liter) and neutralize with 1.5 N HCℓ. Dilute this solution to 500 mℓ to give a stock solution. Prepare a working solution by diluting 5 mℓ of the stock solution to one liter (concentration, 2 μg Mo/liter.

2. Collect the sea water in a nonmetallic sampler and store in a one-liter plastic bottle.

REFERENCES

1. C. R. Goldman, Science, 132, 1016 (1960).
2. R. W. Bachmann and C. R. Goodman, Limnol. Oceanog., 9, 143 (1964).
3. Y. S. Kim and H. Zeitlin, Limnol. Oceanog., 13, 534 (1968).
4. Y. S. Kim and H. Zeitlin, Separation Sci., 6, 505-513 (1971).
5. E. B. Sandell, Colorimetric Determination of Traces of Metals, 3rd ed., Interscience-Wiley, New York, 1959.

Section III

OTHER SPECTROSCOPIC TECHNIQUES

21

PHOTOMETRIC TITRATIONS: CALCIUM AND MAGNESIUM

21.1 Introduction

In preceding chapters, a number of references have been made to coordination entities, or complexes which are formed when metal ions accept a share in the electron pair of a donor atom. These compounds are characterized by their colors and by their persistence in solutions.[1] This property is used in the complexation titration of calcium, magnesium, and even strontium in natural waters. In this titration, a soluble, stable coordination entity of definite composition is formed at the endpoint or equivalence point. The complexation titration, as applied to the determination of calcium and magnesium in natural waters, is not only faster, but can be more precise and more accurate than previous methods, which depended upon precipitation of calcium oxalate.

Probably the most versatile of the complexation reagents is ethylenediaminetetraacetic acid (abbreviated EDTA or H_4Y), which was discovered by Schwarzenbach.[2] There are several advantages to the use of this reagent, including the following: The salt $Na_2EDTA \cdot 2H_2O$ can be obtained in pure form and is a primary standard; the endpoint can be detected in several ways (visually, photometrically, potentiometrically); and the reagent can be used to determine the concentration of several metal ions merely by varying the pH or by adding other reagents.

When EDTA is added to a solution containing both calcium and magnesium, it is possible to determine the concentration of both or of each. Calcium and magnesium are determined together at a pH of 10. When the pH is increased to 12-13, magnesium hydroxide is precipitated, and calcium is determined directly. The magnesium concentration is then obtained by difference. This titration is usually performed visually using an indicator such as murexide and the precision is limited by the sensitivity of the analyst to the color change. With little modification, the complexation titration can be used to determine iron, calcium, and magnesium in limestone.[3]

A photometric procedure has been devised for calcium using murexide,[4] and the accuracy of the calcium determinations is enhanced. When a weight buret is used to add most of the EDTA solution and a microburet is used to add the final increments, calcium can be determined with a precision of 0.005%.[5] This precision is valid for the determination of calcium in pure aqueous solutions. In natural waters, of course, there are considerable amounts of calcium and strontium, and there are three methods of dealing with the problem of interference: correction, separation, or selection.

Pate and Robinson[6] chose the first method. Calcium and magnesium in sea water were determined by an EDTA complexation procedure, and a correction made for the amount of strontium present (cf. Section 22.3). The problem of coprecipitation of calcium with magnesium hydroxide was obviated by adding 95% of the calculated amount of the EDTA solution prior to precipitation of magnesium. Pate and Robinson[6] found the mean Ca/Cℓ ratio (by weight) to be 0.02134 with an average deviation of 0.6%.

Carpenter[5] chose the second method of avoiding interference. Calcium, magnesium, and strontium were separated by adsorption on a cation-exchange resin followed by elution of magnesium (with 0.2 M ammonium acetylacetonate), calcium (with a 1 M solution of the reagent), and strontium (with 2 M ammonium chloride). The separated cations were determined by a photometric titration using EDTA. The method is precise, though not rapid (four duplicate samples can be analyzed in eight hours).

Culkin and Cox[7] chose the third method, titrating calcium selectively in the presence of magnesium using EGTA, a reagent related to EDTA (Note 1). An ammoniacal solution of calcium is treated with an indicator (Zincon, HIn^{3-}) and the zinc complex of EGTA ($Zn\text{-}EGTA^{2-}$) and titrated with EGTA. As the endpoint is reached, an exchange process occurs, Eqn. (21-1), and the solution becomes blue

$$\underset{}{Ca^{2+}} + \underset{\text{(orange)}}{HIn^{3-}} + ZnEGTA^{2-} \rightleftarrows CaEGTA^{2-} + \underset{\text{(blue)}}{ZnIn^{2-}} + H^{+} \quad (21\text{-}1)$$

This procedure is used here (Section 21.2).

21.2 Determination of Calcium by an EGTA Titration[7]

Procedure

(a) Reagents

Buffer Solution. Dissolve 5.35 g of reagent-grade ammonium chloride in 200 mℓ of distilled water, add 7.1 mℓ of concentrated aqueous ammonia (28%, sp. gr. 0.90), and dilute to 250 mℓ.

Zincon Indicator. Dissolve 0.15 g of Zincon in 50 mℓ of methanol. Filter the solution and store at 0-5°.

EGTA Solution (0.005 M). Dissolve 3.80 g of EGTA (Note 1) in 24 m of 1 M sodium hydroxide (Note 2 and dilute to two liters. Standardize against standard calcium solution (Note 3).

Zn-EGTA Solution. This solution must contain exactly equivalent amounts of zinc and EGTA and this is done as follows. Mix 22 mℓ of 0.01 M zinc solution (Note 4) and 20 mℓ of 0.01 M EGTA solution and dilute to 200 mℓ. Titrate 20 mℓ of 0.01 M zinc solution (into to which has been added 5 mℓ of Buffer-I solution and 0.1 mℓ Zincon indicator) with 0.01 EGTA. Next repeat the titration with 10 mℓ of Zn-EGTA solution present. From the difference in volume of 0.01 M EGTA used in the two titrations, calculate the amount of EGTA solution needed to complex the excess of zinc present. Add the appropriate amount of 0.01 M EGTA solution to the Zn-EGTA solution (Note 5).

Standard Calcium Solution (0.01 M). Suspend 1.001 g of primary standard calcium carbonate (Note 6) in 5 mℓ of distilled water and add dropwise a minimum amount of concentrated hydrochloric acid needed to dissolve the solid and dilute to one liter.

(b) Analysis

For each titration, use a sample equivalent to 10 g of sea water of salinity $35^{o}/oo$. Pipet (or weigh) the sample in a 500-mℓ Erlenmeyer flask. Add 5 mℓ of Buffer-I, 1 mℓ of Zincon solution, and 1 mℓ of Zn-EGTA solution. Dilute the solution to 350 mℓ with distilled water and titrate with EGTA solution, using a 25-mℓ buret.

The endpoint can be determined visually (Eqn. 21-1).

A photometric determination is more precise. Place the solution in a photometric titration apparatus (Figure 21-1) (Note 7). Place the tube portion in the cell compartment of a spectrophotometer and cover with a light-proof lid. Thoroughly stir the solution and adjust the spectrophotometer slit so that the instrument reads zero absorbance at 500 mμ.

After each increment of EGTA solution, note the absorbance. When the endpoint is approached (after 21-22 mℓ of EGTA solution), add the titrant in 0.1-mℓ increments, and record absorbance as a function of volume of titrant. After the endpoint occurs, the absorbance remains constant with increasing volume of titrant. Determine the endpoint as the intersection of the straight lines in the titration curve (represented schematically in Figure 21-2).

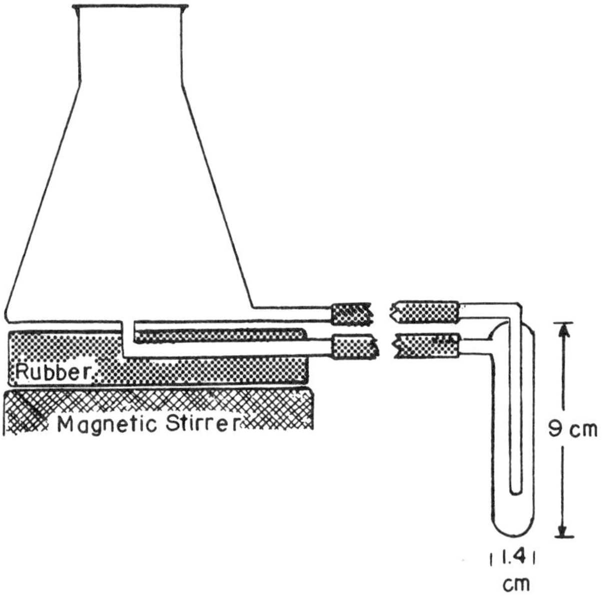

FIG. 21-1

Photometric titration apparatus for use with a colorimeter.

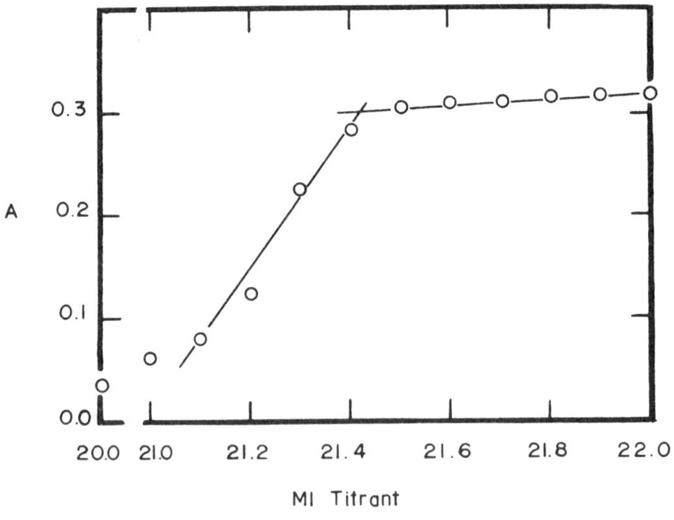

FIG. 21-2

Typical photometric titration plot, using 0.025 M EDTA as titrant and the titration apparatus shown in Figure 21-1.

(c) Calculations

Determine the volume of EGTA solution needed to reach the endpoint. The calcium concentration may be calculated from the following relationships:

$$1\,m\ell\ 0.05\ \text{EGTA solution} = \frac{0.989 \times 10^{-5}}{2}\ \text{g-at Ca (Note 8)} \qquad (21\text{-}2)$$

$$\text{g-at Ca} = \frac{m\ell\ \text{EGTA} \times 0.495 \times 10^{-2}}{m\ell\ \text{sample}} \qquad (21\text{-}3)$$

$$\text{g-at Ca} = \frac{m\ell\ \text{EGTA} \times 0.495 \times 10^{-2}}{\text{g sample}} \qquad (21\text{-}4)$$

$$1\ \text{g-at Ca} = 40.0\ \text{g Ca} \qquad (21\text{-}5)$$

21.3 Determination of Calcium and Magnesium

(a) Reagents

EDTA Solution (0.0254 M). Dissolve 18.610 g of reagent-grade disodium dihydrogen (ethylenedinitrilo)tetraacetate dihydrate (Note 9) in distilled water and dilute to two liters.

Erio Indicator. Dissolve 0.1 g of Eriochrome Black T indicator in 50 mℓ of methanol. Filter the solution and store at 0-5°.

Buffer-II Solution. Dissolve 36 g of reagent-grade ammonium chloride in 300 mℓ of concentrated aqueous ammonia (28%, sp. gr. 0.90) and dilute to 500 mℓ.

(b) Analysis

Pipet or weigh a 10-mℓ sample of sea water in a 500-mℓ Erlenmeyer flask, add 5 mℓ of Buffer-II solution, 1 mℓ of Erio indicator, and dilute with distilled water to 350 mℓ. Transfer the contents of the flask to the photometric titration vessel described in the previous section. Adjust the spectrophotometer slit so the instrument reads zero absorbance at 623 mμ. Record the absorbance as a function of added milliliters of EDTA solution (use 0.1-mℓ increments near the endpoint).

(c) Calculations

Determine the volume of EDTA needed to reach the endpoint (visually or as the intersection of the two linear portions of the photometric titration curve).

From the relationship

$$1 \text{ m}\ell\ 0.025\ \underline{M}\ \text{EDTA solution} = 2.5 \times 10^{-5}\ \text{g-at (Ca + Mg + Sr)} \qquad (21\text{-}6)$$

the total concentration of calcium, magnesium, and strontium may be calculated

$$\frac{\text{g-at (Ca + Mg + Sr)}}{\text{liter}} = \frac{\text{m}\ell\ \text{EDTA}}{\text{m}\ell\ \text{sample}} \times 2.5 \times 10^{-3} \qquad (21\text{-}7)$$

$$\frac{\text{g-at (Ca + Mg + Sr)}}{\text{kg}} = \frac{\text{m}\ell\ \text{EDTA}}{\text{g sample}} \times 2.5 \times 10^{-3} \qquad (21\text{-}8)$$

Subtract the strontium concentration, which can be determined (Chapter 23) or calculated[6,7] from

$$\text{Sr(mg/kg)} = 0.42\ \text{C}\ell^{\text{o}}/\text{oo} \qquad (21\text{-}9)$$

$$\text{Sr(g-at/kg)} = 0.48 \times 10^{-5}\text{C}\ell^{\text{o}}/\text{oo} \qquad (21\text{-}10)$$

Finally, subtract the calcium concentration, as determined by the EGTA titration (Section 21.2) to give the magnesium concentration.

NOTES

1. EGTA: 1,2-bis[2-di(carboxymethyl)amineethoxyl]ethane or ethylene glycol-bis(aminoethyl)tetraacetic acid dry at

50° for an hour prior to use (see Note 9).

2. Prepare by dissolving 4.0 g of reagent-grade sodium hydroxide in distilled water and diluting to 100 mℓ.

3. Use the analysis procedure and take a 10.0-mℓ sample of the standard calcium. This should require exactly 29 mℓ of 0.005 M EGTA solution.

4. Prepare this solution as follows. Dissolve 0.29 g of $ZnSO_4 \cdot 7H_2O$ in distilled water and dilute to 100 mℓ. Dissolve 0.380 g of EGTA in 100 mℓ of distilled water.

5. If the amount of the first titration is 20.00 mℓ and the second titration is 21.80 mℓ, then 1.80 mℓ of 0.01 M EGTA solution must be added to each 10 mℓ of Zn-EGTA solution.

6. G. F. Smith Chemical Co., Columbus, Ohio.

7. A titration apparatus is available from Kontes Glass Co. A titration cell can also be prepared from flat plates of 1/8-in. Lucite sheet. The five plates that form a compartment (3.75 in. deep, 1.75 in. wide, and 4.5 in. long) are attached to each other by softening with ethylene dichloride. Other cells have been described by Carpenter[5] and Headridge.[8] (See Figure 21-1.)

8. The factor 0.989 arises from the observation that magnesium and strontium (in the amounts present in sea water) affect the titre of the EGTA solution. The interference is minimal at low Mg:Ca ratios.

9. $Na_2EDTA \cdot 2H_2O$: (Ethylenedinitrilo)tetraacetic acid disodium salt, dihydrate. The reagent-grade material should

be dried at 50° for an hour before weighing out. It is a primary standard,[3] but after a few weeks in glass, the concentration will change. Store in a plastic bottle and check the concentration against standard calcium solution periodically. The material dissolves slowly.

REFERENCES

1. D. F. Martin and B. B. Martin, Coordination Compounds, McGraw-Hill, New York, 1964.

2. G. Schwarzenbach, Complexometric Titrations, Methuen and Co., Ltd., London, 1957.

3. W. J. Blaedel and V. W. Meloche, Elementary Quantitative Analysis, 2nd ed., Harper and Row, New York, 1963, pp. 590-599.

4. M. B. Williams and J. H. Moser, Anal. Chem., 25, 1414 (1953).

5. J. H. Carpenter, Limnol. Oceanogr., 2, 271 (1957).

6. J. B. Pate and R. J. Robinson, J. Mar. Sci., 17, 390 (1958).

7. F. Culkin and R. A. Cox, Deep-Sea Res., 13, 789 (1966).

8. J. B. Headridge, Photometric Titrations, Pergamon Press, Oxford, 1961.

22

ATOMIC ABSORPTION SPECTROSCOPY

22.1 Introduction

Atomic absorption spectroscopy (AAS) is a relatively new method for determining the presence and amount of an element in a sample. Walsh realized the potential application of the method to the analysis of routine samples[1] and developed an apparatus which was suitably simple, versatile, and inexpensive. The method appears to be an excellent one for determining the concentration of certain trace metals in natural waters because of its rapidity, sensitivity, and simplicity. It has few chemical or spectral interferences, and those chemical separations that may be required are simple ones. The concentrations of many trace metals in sea water are less than the detection limits of AAS, and some extraction procedure must be used, contrary to the impression gained from early reports.[2]

This chapter should provide an introduction to the applications and limitations of the method. Robinson[3] and Kahn[4,5] have provided a more extensive coverage of atomic absorption spectroscopy.

Flame photometry and AAS resemble each other in the superficial sense that a sample is aspirated into a flame and atomized. Significant differences exist between the two methods, however, that may become clear when subsequent events are considered. When sufficient radiant energy is supplied to free atoms, some valence electrons (those outermost electrons that are responsible for chemical behavior) will absorb characteristic amount of energy and will be converted from a "ground" state to an "excited" state or higher energy level. The higher energy level is unstable, and the electron will release a characteristic amount of energy and return to the original ground state.

These events provide a basis for the contrast between the two methods. Flame photometry measures the intensity of light emitted by excited species at an appropriate wavelength and relates intensity to concentration of element in the sample by means of calibration curves. In AAS, however, the atom is merely dissociated from chemical bonds by means of a flame, and is converted to an unexcited, un-ionized atom in the ground state. In addition, in AAS, a light beam is sent through the flame and (after being sent into a monochromator and a detector) the amount of light absorbed is measured and related to concentration. Absorption is often more sensitive than emission since it depends upon the presence of free, unexcited atoms, and the ratio of these to excited ones is usually large at a given instant.

The principle of AAS is simply stated. The wavelength of light from a cathode lamp is characteristic of the element being analyzed and the energy absorbed in the flame is related to the concentration of the element in the sample.

22.2 AAS Equipment Components

Basically, atomic absorption equipment consists of four units: a radiant-energy source, a sample cell, a monochromator, and a detector. The arrangement of these units is shown schematically in Figure 22-1.

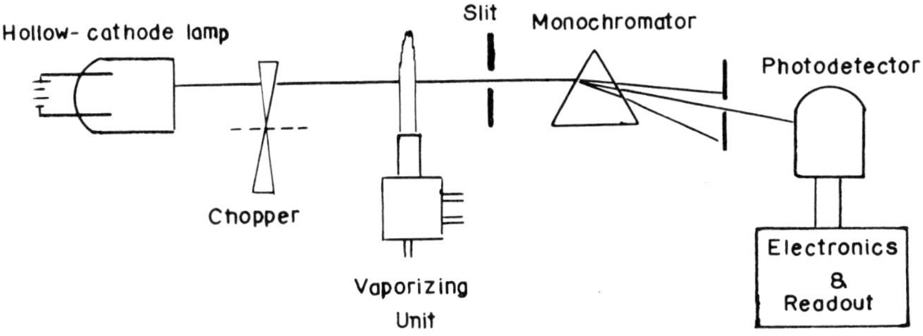

FIG. 22-1

Schematic representation of an atomic absorption system.

The radiant-energy source is light from a cathode which is made of the element being measured. This light of intensity I_0 at a given wavelength is passed through a sample which has been vaporized in a flame. Ground-state atoms in the flame absorb the light, and the intensity is reduced to a value I_1. In addition, light from the flame is emitted in all directions; this light has intensity S at the given wavelength. The extra light reduces the sensitivity and must be corrected for as much as possible. The per cent absorption is noted in a detector and is taken as a measure of the concentration of the element in the sample.

The atomic absorption process is represented schematically in Figure 22-2.[4] The emission spectrum of the hollow-cathode

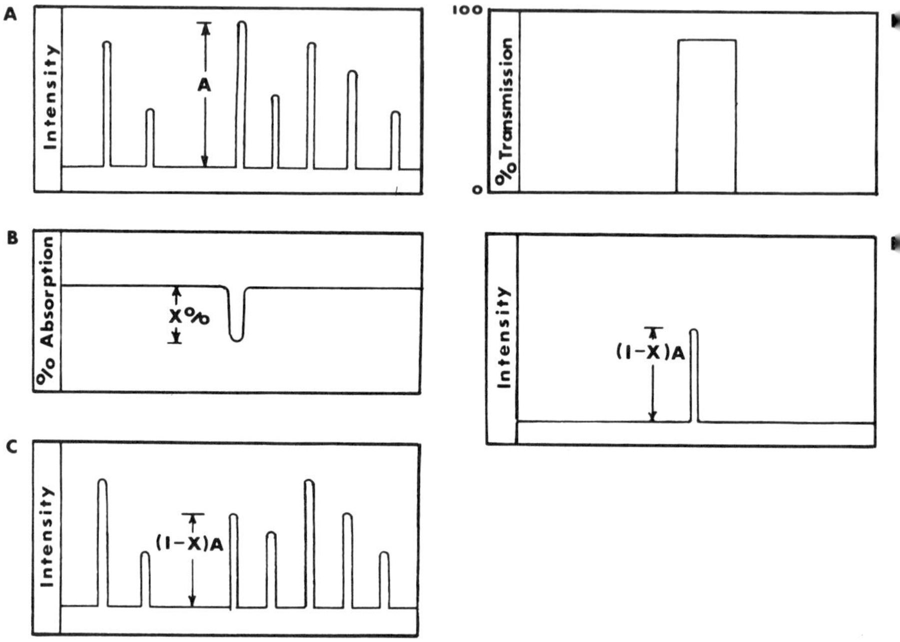

FIG. 22-2
Atomic absorption process (from Kahn,[4] redrawn with permission).

lamp (Figure 22-2A) has typically narrow lines with a half-width of less than 0.01 Å). The element of interest in the sample can absorb only energy of a certain wavelength, which corresponds to the so-called "resonance" line. The element of interest absorbs some amount of the resonance line (Figure 22-2B); the amount x corresponds to the concentration of the element. The resultant spectrum (Figure 22-2C) shows that the intensity of the resonance line is reduced; all other lines are unaffected. The monochromator or filter is set up to pass only the resonance line (Figure 22-2D). The photodetector then sees only the resonance line, diminished by absorption.

22.3 Advantages of AAS

The advantages of speed, sensitivity, and need for only a small sample have been described. Other advantages include the following.

Accuracy and Precision. High precision and accuracy are feasible with the aid of calibration curves. These curves tend to be more valid than with other analytical methods because of the general absence of interference.

Limited Interferences. Interference by atoms of other elements is rare because each metal absorbs at a well-defined and relatively narrow wavelength range. For example, isotopes of the same element may be determined because the isotopes do not absorb each other's radiation.

Because of limited interferences, separations are not needed and the possibility of contamination by trace impurities is obviated.

Chemical interference can arise, however, when stable compounds cannot be broken down and the neutral metal atom cannot be produced readily. The decomposition of chemical compounds is a problem common to most analytical processes.

Simplicity. Comparatively little training is needed to do the analyses. Commercial instruments are simple to operate, and with a minimum of training, the operator can perform routine analyses with a good degree of precision and accuracy (Table 22-1).

TABLE 22-1

Atomic Absorption Detection Limits of Selected Elements[a]

Element	Detection Limit, ppm	Wavelength, mμ
Aluminum[b]	0.1	309.3
Barium[b]	0.05	553.6
Cadmium[c]	0.001	228.8
Chromium	0.005	357.9
Cobalt	0.005	240.7
Copper	0.005	324.7
Iron	0.07	248.3
Lead	0.03	283.3
Lithium	0.005	670.8
Manganese	0.002	279.5
Nickel	0.005	232.0
Silver	0.005	328.1
Vanadium[b]	0.02	318.4
Zinc	0.002	213.8

[a]Ref. 5. Perkin-Elmer Model 303 double-beam Atomic Absorption Spectrophotometer using aqueous solutions.

[b]Nitrous oxide flame.

[c]Argon-hydrogen flame.

22.4 Problems in AAS

Suitable Elements. Over 65 elements have been detected by atomic absorption.[5] Few methods have been developed for detecting nonmetals. This is because the absorption lines are in an unfavorable region, e.g., the vacuum-ultraviolet. Determination of phosphorus, sulfur, and chlorine by atomic absorption involve preliminary conversion to a metal compound and analysis of the compound.[6]

Oxide Formation. The use of flame atomizers in most commercial equipment limits the sample type to liquids. For many samples, this is satisfactory, but some metals form stable oxides in the flame, with resultant loss in sensitivity. This is a problem with aluminum, beryllium, vanadium, and titanium, and a high-temperature flame, e.g., nitrous oxide, must be used.

Simultaneous Analysis. This is generally not feasible on most commercial instruments, though the limitation is not serious unless only a very small sample is available.

Anion Interference. In some instances, anion interference due to the formation of unusually stable metal-anion bonds must be overcome. A good example is anion interference by chloride, phosphate, carbonate, iodide, sulfate, and fluoride in the determination of lead. In this instance, the problem is overcome by adding 1% EDTA solution.

22.5 Light Source and Sample Vaporization

As noted before, there are four basic units of the atomic

absorption spectrophotometer: the light source, a device for vaporizing the sample, a system for isolating the resonance line, and a detector. These are considered in detail by Robinson[3] and by Kahn,[4,5] but the following is a summary of two important features, the light source and vaporization of the sample.

Light Source. The most convenient source is the spectral vapor lamp, which can be used for the more volatile elements (the alkali metals, mercury, and thallium). The most common source is the hollow cathode discharge lamp, which is used for the less volatile elements. The hollow cathode is a more successful source than the first type or continuous source (such as a hydrogen lamp) because of the narrowness of the atomic absorption lines.

The single-element hollow cathode lamp is available based on a design by Jones and Walsh. A voltage applied between the anode and cathode causes atoms of helium or argon to be charged at the anode and to be drawn toward the cathode. The charged particles bombard the metallic cathode and dislodge excited metal atoms into the gas-filled tube. Excited metal atoms emit light of wavelengths which are characteristic of that metal. This results in a characteristic light as well as a cloud of metal atoms, which diffuses out or redeposits on the cathode.

Sample Vaporization. Atomic absorption spectroscopy has been applied almost exclusively to liquid samples, and the flames used are similar to those used for flame photometry (Note 3).

For routine work, an acetylene-air mixture (2000-2200°) is probably superior to other mixtures from the standpoint of convenience and sensitivity. Other mixtures and the approximate flame temperatures are: nitrous oxide-acetylene (3000°), hydrogen-air (2100°), acetylene-oxygen (3100°), propane-oxygen (2700-2800°), illuminating gas-oxygen (2800°).

Commercially available atomizers are burners, and two types of burners are the "total consumption" type and the "premix" type. The first type is so named because all the sample that is aspirated is injected into the flame. With this burner, a representative sample reaches the flame, and there is no explosive hazard from unburned gas mixture. It has several disadvantages: The sample aspiration depends on the viscosity of the liquid, and some viscous liquids cannot be aspirated. The size of the drops injected into the base of the flame may vary widely even under optimum conditions. The feed rate may vary because of encrustation in the burner tip or flow lines and the absorption signal can vary significantly. Finally, the burner is noisy (both electronically and physically).

Some of these disadvantages are absent in the premix types. Oxygen and fuel are premixed in a sample vaporizing chamber, the sample is aspirated into the same chamber, and a sizable fraction is evaporated into the fuel-oxidant mixture. The whole mixture is swept into the flame, where combustion occurs. A portion of the aspirated sample is not combusted and drips out through an excess sample drain. The premix-type burner is much less noisy, is less susceptible to encrustation, and is able to introduce more atoms into the light path, all of which improve the sensitivity of the method.

The premix burner has two major disadvantages: Serious errors in accuracy may arise if mixed solvents are used, because there is a tendency for the more volatile solvent to be vaporized selectively; some metal compounds can thus remain behind in the less volatile fraction. Also, the burner uses relatively large volumes of fuel-oxidant mixtures, and the danger of explosion is increased; some mixtures cannot be used for this reason.

22.6 Determination of Selected "Soluble" Metals by AAS

The concentration of a soluble species of trace metal element in sea water is often less than the detection limits of AAS, and this convenient technique cannot be used unless some means of concentration is found. The procedure described here is based on a solution to the problem devised by Brooks and co-workers.[6] Various trace metals are complexed with ammonium pyrrolidine dithiocarbamate (APDC), and the resulting coordination entities are extracted into methyl isobutyl ketone (MIBK) and analyzed by AAS.

The procedure has several useful features. The complexing reagent, APDC, is insoluble in MIBK, which provides a means of purification of the reagent and, subsequently, a means of separation of the coordination entities from the excess complexing agent. A large sea water/solvent ratio is used to achieve a favorable distribution coefficient and enhance the concentration of trace metal. Allowance is made for incomplete extraction of the trace metal being determined. Finally, particulate matter is analyzed simultaneously by dissolving membrane filters in an acetone-hydrochloric acid mixture. The specific instrumental and analytical parameters are listed in Table 22-2.

TABLE 22-2

Instrumental and Analytical Parameters

Element	Analysis wavelength, mμ	Detection limit, ppb	Coefficient of variation, %
Cobalt	241	0.10	15
Copper	325	0.10	4
Iron	248	0.10	6
Lead	317	0.20	8
Nickel	232	0.10	11
Zinc	214	0.05	3

Procedure

(a) Reagents

Deionized Water. See Section 8.4.

Methyl Isobutyl Ketone (MIBK). Redistill commercial-grade material.

Ammonium Pyrrolidine Dithiocarbamate (APDC). Prepare daily by transferring 2.0 g of APDC to a separatory funnel containing about 200 mℓ of deionized water. Shake to dissolve as much as possible, add 200 mℓ of redistilled MIBK, and shake again. Run the lower aqueous layer through fluted filter paper and discard the first 20 mℓ. Store the remainder in a polyethylene bottle.

Hydrochloric Acid (6 N). Dilute high-purity, concentrated hydrochloric with an equal volume of deionized water.

Stock Metal Solutions. The stock solution should contain 38.5 ppm each of iron, copper, and zinc (Note 1). Dissolve separately 38.5 mg of electrolytic iron (Section 19.2) in the minimum volume of concentrated hydrochloric acid, then 38.5 mg of copper in the minimum volume of concentrated hydrochloric acid (Section 17.2), and finally 38.5 mg of pure zinc in the minimum volume of concentrated hydrochloric acid. Carefully wash the three solutions into a one-liter volumetric flask with deionized water and dilute to nearly one liter. Adjust the pH to 4 with concentrated hydrochloric acid.

Working Standard (1.54 ppm). Prepare daily by diluting 10.0 mℓ of stock standard to 250 mℓ with deionized water.

Seawater Standards. The standards contain 0, 2, 5, 10 ppb metal and are prepared by adding to 750 mℓ of extracted seawater 0, 1.0, 2.5, and 5.0 mℓ of working standard, respectively.

(c) Analysis ("Soluble Metals")

Filter one-liter samples of sea water through a 0.45 μ membrane filter, and save the filter for particulate-metal analysis. Adjust the pH to 4-5 with hydrochloric acid. The step is necessary to prevent adsorption or precipitation during prolonged storage, but may be omitted if analysis begins within a short interval after collection.

Complex the trace metals and separate the complexes as follows. Place 750-mℓ samples of the sea water (measured with a graduated cylinder) in polyethylene Erlenmeyer flasks and add 35 mℓ of redistilled MIBK and 7 mℓ of APDC solution to each flask. Equilibrate by shaking on a mechanical shaker for 30 minutes, then separate the phases by means of a separatory funnel. The organic phases can be saved by storing in a polyethylene bottle (Note 2). The aqueous layer should be saved for use in preparing calibration curves.

Analyze the MIBK extracts by aspirating into a fuel-lean air-acetylene flame, using the wavelengths given in Table 22-2.

Prepare a calibration curve by using the extracted seawater samples. Add 20 mℓ of MIBK to each of eight samples of extracted aqueous layer and equilibrate for five minutes by shaking. Separate the aqueous phases and recombine all eight. This will ensure homogeneity. Treat each 750-mℓ sample in an Erlenmeyer flask with enough working standard solution to give the four concentrations described under "seawater standards". Measure the temperature of the water, and add enough MIBK to compensate for the amount of ketone dissolved after equilibration (Note 3). Then, add 7 mℓ of APDC solution, shake for 30 minutes, and separate phases as before. The zero point of the calibration curve is usually adequate as a blank.

(d) Analysis ("Particulate Metals")

Extract the particulate metals from the membrane filter. Wash the filter with 50 mℓ of deionized water and place it in a polyethylene bottle. Add 5 mℓ of 6 N hydrochloric acid, and seal the bottle in a 70° bath for one hour. Add 15 mℓ of acetone to dissolve the membrane filter.

Analyze the solution as before against suitable standards.

NOTES

1. If desired, standards can be prepared containing 38.5 mg of Co (143 mg $CoBr_2$) and Pb (51.7 mg $PbCl_2$).
2. Most complexes of APDC are stable for about three hours; the cobalt and copper complexes are stable for about 24 hours. The extracts can be preserved for a long time by freezing in a Dry Ice-acetone bath.
3. The use of a flameless atomic absorption technique that employs a graphite atomizer reduces the detection limits by factors of 0.1-0.01 and may eliminate the need for preconcentration.[7]
4. The volumes to be added are 7 mℓ (30°), 6.5 mℓ (25°), 6 mℓ (20°).

REFERENCES

1. A. Walsh, Spectrochim. Acta, 7, 108 (1955).
2. B. P. Fabricand, R. R. Sawyer, S. G. Ungar, and S. Adler, Geochim. Cosmochim. Acta, 26, 1023 (1962).
3. J. W. Robinson, Atomic Absorption Spectroscopy, Dekker, New York, 1966.
4. H. L. Kahn, J. Chem. Educ., 43, A7/-A107 (1966).
5. H. L. Kahn, Adv. Chem. Ser., 73, 183 (1968).
6. R. R. Brooks, B. J. Presley, and I. R. Kaplan, Talanta, 14, 809 (1967).
7. D. C. Manning and F. Fernandez, Atomic Absorption Newsletter, 9(3), 65-70 (1970).

23

FLAME PHOTOMETRY: CALCIUM AND STRONTIUM

23.1 Introduction

Flame photometry is a useful method for determining the presence and amounts of certain alkali metals (lithium, sodium, and potassium) and alkaline-earth metals (magnesium, calcium, and strontium) in natural waters. For these and some other elements, flame photometry is to be preferred to atomic absorption spectroscopy (Chapter 22). The sensitivities of the two methods for various elements are compared in Table 23-1.

There are six major components of a flame photometer. These are represented schematically in Figure 23-1. The components include: the pressure regulators and flow meters for the fuel gases, the atomizer, the burner, the optical system,

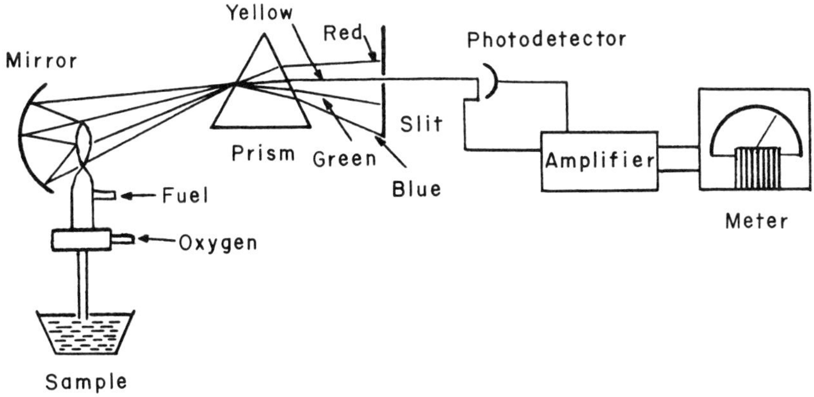

FIG. 23-1

Schematic representation of a flame photometry system (from Dean,[1] redrawn with permission).

TABLE 23-1

Comparison of Atomic Absorption and Flame Photometry Sensitivity[1,9]

Element	Flame Photometry, ppm/%T		Atomic Absorption, ppm
Aluminum	0.4	(484)[a]	0.8
Barium	3	(455.5)	3.5
	4	(493.4)	0.2
Boron	3	(318)	—
	3	(546)	—
Cadmium	0.5	(326.1)	0.01
Calcium	0.07	(422.7)	—
	0.16	(554)	0.08
	0.6	(622)	—
Cesium	2.0	(455)	0.15
Chromium	5	(425.4)	0.006
Cobalt	3.4	(345.4)	0.013
Copper	0.6	(324.7)	0.005
Iron	2.7	(386)	0.05-0.1
Lead	14	(405.8)	0.013
Lithium	0.067	(670.8)	0.03
Magnesium	1.4	(371)	0.001
Manganese	0.1	(403.3)	0.005
Nickel	1.6	(352.4)	0.01
Potassium	0.02	(767)	0.03
Silver	0.6	(338.3)	0.05
Sodium	0.001	(589)	0.03
Strontium	0.06	(460.7)	0.05
Vanadium	12	(523)	0.5
Zinc	500	(213.9)	0.005

[a]Wavelength, mμ.

the detector, and the meter which indicates the output of the detector. A detailed description of these components is given by other authors[1,2] or in the literature accompanying a given instrument.

The function of each component will become apparent from a brief consideration of the flame photometry experiment. A solution of the element to be detected is injected into a very hot flame (2000°C or greater). The characteristics of the flame are controlled to the extent possible by the use of different fuels and by standardizing the flow rate of the gases. The thermal energy of the flame causes valence electrons to be excited from the ground-state level to a higher energy level or orbital. This is not a stable situation and the electron returns to the lower energy level and in doing so, emits light of a characteristic wavelength. The wavelength of the emitted light is characteristic of the element, and the intensity of the light is characteristic of the amount of element present. The excited electron may not return to the initial energy level directly, but in a series of steps; therefore, several emission wavelengths will be observed.

Obviously, the wavelength and intensity of emitted light must be distinguished from that of the background light. This is done with the aid of several devices. The collimating mirror may be so placed that its center of curvature is at the flame and the intensity of the flame emission is doubled. Next, the optical system collects light from the steadiest portion of the flame. The light is made monochromatic by means of a prism, a grating, or filters and is focused on the photodetector, typically a photomultiplier tube. The signal from the photomultiplier tube is increased in the amplifier, which also aids resolution of close spectral wavelengths and which aids the identification of elements present in trace concentrations.

In the sections that follow, the major interferences and the techniques of flame photometry will be considered. The procedure for determining calcium and strontium will be given as an example of the method.

23.2 Spectroscopic Interferences

There are two categories of interference, spectroscopic and sample. Interferences of the first type include the following:

Stray Radiation. The detector may pick up light from elements other than the one being measured. This problem is governed mainly by the resolution of the instrument and the slit width used. Even so, the problem may be serious, depending upon the circumstances. For example, potassium and manganese emission occurs at 404.4 and 403.3 mμ, respectively, and the intensity of the manganese emission is much greater. The interference of trace amounts of potassium on manganese would be slight compared with the marked interference of trace amounts of manganese in potassium analyses.

Reproducibility of the Flame. The energy of emission is

a function of the constancy and temperature of the flame. In some instances, the temperature of the flame may be a limiting factor in determining whether or not the metal can be detected. Optimum excitation conditions are obtained with an acetylene-oxygen flame, which can be used in the detection of some 45 elements. The quality and composition of the fuel affect the properties of the flame: the constancy and temperature, the shape, background, and rate of sample consumption.

Combustion Products. These may interfere by affecting the flame or by affecting the optical system (fogging or coating the lenses and mirrors).

Flame Background. The reading obtained from the detector includes a contribution from the flame background upon which the emission line is superimposed. A serious error would result if this background were not corrected. The background radiation is read directly in the presence of the test element when the flame photometer has a good monochromator.

This correction is made as follows. Measure, in the usual way, the intensity of the emission (line plus background) at the emission line of appropriate wavelength, i.e., the one of maximum intensity. Next, rotate the wavelength dial to the left or to the right until the emission reading reaches a minimum (background). Do the same on the other side of the emission-line wavelength. If the intensity readings are the same to the right and to the left of the emission line, merely subtract the intensity of the background from the intensity of the line-plus-background reading. If the intensities of background differ, interpolate to the value at the emission line.

23.3 Sample Interferences

Interferences of this type include the following.

Self-Absorbing Elements. At some concentrations, the energy of excitation is not proportional to the concentration of the element being detected. For example, for the emission line of sodium at 589 mμ, the per cent transmittance is not a linear function of the sodium concentration above 10 ppm. At greater concentrations, self-absorption effects become more prominent. This problem is overcome by diluting the sample to permit reading on the linear portion of the calibration curve.

Self-absorption arises when the emitted radiation (from excited ion returning to the ground state) is absorbed by non-excited ions of the same element. The intensity at the pertinent wavelength is reduced.

Elements Emitting at the Same Wavelength. Two elements, such as calcium and chromium, may produce light at the same wavelength (423 mμ and 425 mμ). The two elements interfere by an additive effect. Either a different wavelength must be selected or the unwanted element must be masked or removed.

Radiation Interference. One element may cause another element to modify its normal emission intensity, either in a positive or negative manner. The problem is corrected either by dilution or controlled interference addition. For example, sodium may enhance the emission of small amounts of potassium.

Anion Interference. Problems of anion interference in atomic absorption spectroscopy are also found in flame photometry. For example, phosphate can lower the emission intensity of calcium, possibly because of the formation and stability of calcium pyrophosphate in the flame.[3] This problem is overcome by adding disodium ethylenediaminetetraacetate in strongly alkaline solution, and the calcium is converted to a complex that decomposes in the flame.[4]

23.4 Representative Techniques

The following techniques are typical of procedures currently used to overcome interferences in routine and in special samples.

Emission Intensity As a Function of Concentration. When no interference is present, a standard calibration curve (per cent versus concentration, milligrams/liter) is prepared from solutions containing known concentrations of the element to be determined. The emission intensity of the unknown sample is determined, and the concentration is obtained from the calibration curve.

When an interfering element is present, it must be compensated for or removed. The latter is time-consuming and added reagents may also introduce impurities. Three compensation methods follow.

Radiation Buffers. This method depends upon the fact that radiation interference reaches a maximum for large amounts of interfering elements. Thus for measurements of sodium, potassium, calcium, and magnesium, radiation buffers are prepared as solutions saturated with respect to each metal. For example, for the determination of sodium in river water, a sodium buffer was prepared.[5] This buffer was prepared by saturating distilled water with potassium chloride, calcium chloride, and magnesium chloride. The sodium buffer (about 1 mℓ) was added to a known volume of test sample and the sodium emission intensity of the sample was determined.

Internal-Standard Method. To each sample and standard, a fixed amount of internal standard element (lithium) is added to reduce effects of variations. In burner and atomizer operations, the internal standard element must be one that is not already present in the sample. Typically, lithium stock solution (5000 ppm, 49.85 g of lithium nitrate) for internal standard measurement is prepared in large volume at ten times the strength to be used in the sample. In the standard and unknown solutions, the lithium concentration should be 500 ppm, which appears to give the best results from a joint precision and sensitivity standpoint when the sodium and potassium concentrations are less than two milliequivalents per liter.

The emission intensities of internal standards and test elements are read simultaneously or separately, depending on the instrument.

A correction is made for background intensity. A plot of the log emission intensity rate (test element/internal standard) as a function of the log of the concentration of test element should be linear with a slope of about 45° at suitable concentrations.

This method has been applied to analysis of biological samples.[6,7]

Standard-Addition Method. This is, in effect, a modification of the previous method with the internal standard a known concentration of test element. The determination of calcium in sea water and marine organisms, as developed by Chow and Thompson,[8] is used here as a good example of the method.

To known volumes of the test solution are added equal volumes of series of five standard solutions containing a known amount of the test element, in this case, calcium. All samples are diluted to the same volume. The emission intensities of these solutions are determined at the appropriate wavelength and a correction is made for flame background. The net emission as per cent transmission is plotted as a function of the standard concentration in the unknown solutions (cf. Fig. 23-2).

The concentration of the unknown is determined as follows. The per cent transmission of the mixture containing unknown plus the zero standard (distilled water) is equal to b. From the graph, determine the concentration x for which the intensity is equal to double this value, 2b. This is the concentration in the original sample (Notes 1,2).

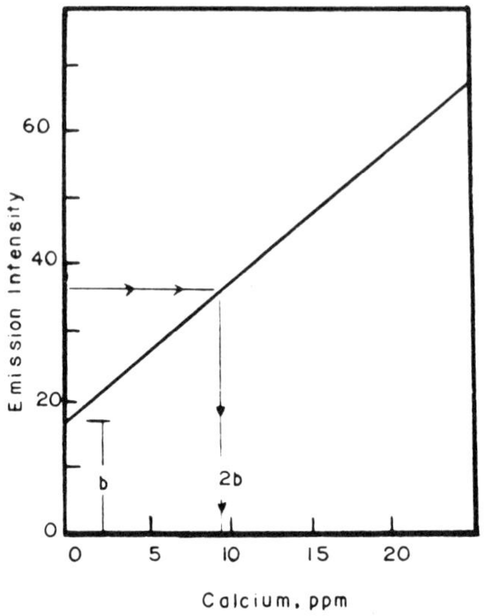

FIG. 23-2

Determination of calcium by an internal standard procedure (after Chow and Thompson[10]).

23.5 Calcium Analysis

Procedure

(a) Reagents

Concentrated Hydrochloric Acid (12 M). Use 37% acid (sp. gr. 1.19).

Distilled Water. Ordinary distilled water should be deionized (Section 8.4) and stored in a polypropylene container.

Standard Calcium Solution. Carefully dissolve 1.001 g of reagent-grade calcium carbonate in 5 mℓ of concentrated (12 M) hydrochloric acid and dilute to one liter with distilled water. This stock solution contains 10 mg-at Ca/liter. Prepare five "internal" standard solutions by pipeting 100.0, 80.0, 60.0, 40.0, and 20.0 mℓ of the stock solution in one-liter volumetric flasks and dilute to one liter with distilled water. The five internal standard solutions and the stock solution should be stored in polypropylene bottles to prevent contamination with sodium ion. Concentration data are given in Table 23-2. The zero standard is prepared by diluting 5 mℓ of concentrated (12 M) hydrochloric acid to one liter with distilled water.

(b) Instrument Parameters

The following instrument parameters have been used.[8]

Wavelength for calcium spectral line	422.7 mμ
Wavelength for the flame background	418 mμ
Selector switch setting	1.0
Slit width	0.001
Fuel gases	
Oxygen	15 psi
Hydrogen	4 psi
Sensitivity	clockwise
Phototube	multiplierphototube
Resistor	22 megohms

Obviously, some of these parameters are a variant (e.g., wavelengths) and others are subject to adjustment or will depend upon the particular instrument used.

TABLE 23-2

Calcium Analysis Standard Addition Data

Standard solution No.	0	1	2	3	4	5
Milliliters of stock solution per liter	0	20.0	40.0	60.0	80.0	100.0
Standards concentration						
mg Ca/liter	0	8.0	16.0	24.0	32.0	40.0
mg-at Ca/liter	0	0.20	0.40	0.60	0.80	1.00
Standards concentration (in test solution)						
mg Ca/liter	0	4.0	8.0	12.0	16.0	20.0
mg-at Ca/liter	0	0.10	0.20	0.30	0.40	0.50

(c) Analysis of Natural Waters

Sea Water. Dilute 20 mℓ of the seawater sample to one liter and mix thoroughly. Pipet 5 mℓ of the sample into each of six tubes. To the first tube, add 5 mℓ of distilled water, and mix. To the other five tubes, add 5 mℓ of one of the five internal standards. The concentration of the internal standard in each of these five tubes is given at the bottom of Table 23-2.

Determine the emission intensity (per cent transmission scale) following the instructions that accompany the instrument and using the instrument parameters given before. Determine the net light intensity emitted by calcium by subtracting the reading obtained at 418 mμ (flame background) from the reading obtained at 422.7 mμ (emission plus flame background). Rinse the atomizer-burner thoroughly with distilled water before determining the next emission of the sample. Continue until the emission intensity of all six test samples has been determined.

Fresh Water. There is no need to dilute the sample, and six 5-mℓ portions may be taken directly from the sample.

(d) Analysis of Solid Samples

Accurately weigh a 0.50-1.0 g sample which has been dried at 105° overnight. Ignite the dried material to 500° in order to destroy organic material. Allow the sample to cool. Add 10 mℓ of distilled water. Then, add concentrated (12 M) hydrochloric acid dropwise until the residue dissolves. Carefully transfer the contents to a one-liter flask and dilute to one liter with distilled water. Use 5-mℓ portions and carry out the analysis as described previously for natural waters.

(e) Calculations

Plot the emission intensity, corrected for flame background, for each sample, as per cent transmission versus concentration of calcium standard (Table 23-2, Fig. 23-2). Double the emission intensity or the zero standard. From the value (= 2b), read off the value x, the observed calcium concentration.

The concentration of calcium in the original unknown sample is

$$50x \text{ mg Ca/liter (for seawater samples)} \qquad (23\text{-}1)$$

$$x \text{ mg Ca/liter (for freshwater samples)} \qquad (23\text{-}2)$$

$$\frac{0.1x}{\text{grams of dried sample}} \text{ \% Ca (for solid samples)} \qquad (23\text{-}3)$$

23.6 Strontium Analysis

Procedure

(a) Reagents (Note 3)

Standard Strontium Solution. Carefully dissolve 0.739 g of strontium carbonate in 5 mℓ of concentrated (12 M) hydrochloric acid and dilute to one liter with distilled water. This stock solution contains 5 mg-at Sr/liter or 438 mg Sr/liter. Prepare five internal standards using the same general procedure as with the calcium standards (Table 23-3).

TABLE 23-3

Strontium Analysis Standards Data

Standard solution No.	0	1	2	3	4	5
Milliliters of stock solution per liter	0	10.0	20.0	30.0	40.0	50.0
Standards concentration						
mg Sr/liter	0	4.38	8.76	13.14	17.52	21.90
mg-at Sr/liter	0	0.05	0.10	0.15	0.20	0.50
Standards concentration (in test solution)						
mg Sr/liter	0	2.19	4.38	6.57	8.76	10.95
mg-at Sr/liter	0	0.025	0.03	0.075	0.10	0.25

(b) Instrument Parameters (Note 4)

The following instrument parameters have been used.[10]

Wavelength for the strontium spectral line 460.7 mμ

Wavelength for the flame background 454 mμ

Selector switch setting 0.1

Slit width	0.02 mm
Fuel gases	
Oxygen	15 psi
Hydrogen	4 psi

(c) Analysis and Calculations

The procedures are the same as those used for the analysis of calcium.

NOTES

1. The straight line in the figure is given by the relationship %T = mx + b, where m is the slope; x is the concentration of calcium in the unknown sample; b is the intercepter per cent transmission for the zero standard sample. By definition, b = (x/2 + 0/2), or 2b = x.
2. The original sample may have been diluted before mixing with standard. If so, the observed concentration must be multiplied by an appropriate factor to give the true concentration of the original sample. When 20 mℓ of sea water is diluted to a liter, the factor is 50.
3. All solutions should be stored in polypropylene bottles.
4. Other instrument parameters are those used for the calcium determination.

REFERENCES

1. J. A. Dean, Flame Photometry, McGraw-Hill, New York, 1960.
2. R. Herman and C. T. J. Alkemade, Chemical Analysis by Flame Photometry, Interscience, New York, 1963.

3. G. L. Baker and L. H. Johnson, Anal. Chem., 26, 465 (1954).

4. J. D. Wirtschafter, Science, 125, 603 (1957).

5. P. W. West, P. Folse, and D. Montgomery, Anal. Chem., 22, 667 (1950).

6. C. L. Fox, Jr., Anal. Chem., 23, 137 (1951).

7. W. M. Wallace, M. Holliday, M. Cushman, and J. R. Elkinton, J. Lab. Clinical Med., 37, 621 (1951).

8. T. J. Chow and T. G. Thompson, Anal. Chem., 27, 910 (1955).

9. J. W. Robinson, Atomic Absorption Spectroscopy, Dekker, New York, 1966.

10. T. J. Chow and T. G. Thompson, Anal. Chem., 27, 18 (1955).

Section IV

ANALYSIS OF POLLUTED WATERS

24

ANALYSIS OF POLLUTED WATERS

24.1 Introduction

At first glance, the topic of pollution may seem to be inappropriate to marine chemistry, because of the total volume of the world's oceans. As Ryther noted,[1] considered as a whole, the oceans could serve as a sewage treatment plant for mankind for many years. As a supporting example, assume the waste from six billion persons (an estimated world population in 30 years) were transported to the oceans and uniformly and rapidly mixed. These wastes would not exhaust the dissolved oxygen content of the world's oceans in less than 70,000 years (even with no regeneration due to photosynthesis) and combined inorganic nitrogen would barely double in 100,000 years.

Unfortunately, the qualifying conditions of "rapid mixing and uniform distribution" do not occur and discharge of waste is highly localized and concentrated. The pollution of estuaries seems to be generally recognized;[2,3] while pollution of coastal waters is increasingly obvious, the factors responsible for coastal and marine pollution are manifold, including substantial population increases, with concomitant increases in manufacturing capacity and industrial wastes, and proximity of the population to the coastlines (30% of the U. S. population is within 50 miles of coastlines). As a consequence, marine waters will increasingly become the final repository of wastes.

At present, there is a paucity of scientific information concerning the effects of pollutants on economically valuable marine life in nonrestricted waters. Reaching conclusions is a difficult task in view of the myriad kinds of wastes and the vast number of effects to be considered. Some of the difficulties will become apparent in the next section.

24.2 Pollution Parameters

(a) Sampling Costs

These must surely vary with the location and the nature of the problem. Accurate sampling seems to be a particularly difficult task in a polluted estuary, such as the Delaware River estuary, because of the complexities of movement and buildup of pollutants.[4]

A marine monitoring program in California reported[5] the cost estimate approximated the formula: ℓog Cost = 0.8 ℓog Quantity + 2.6, where Cost is the total annual (dollar) cost of sampling and Quantity is the volume of discharge (million gallons per day, mgd). The City of Los Angeles and Los Angeles Sanitation Districts discharged more than 200 mgd in December 1961. Monitoring costs increase rapidly when the discharge volume is less than 1 mgd, but, by the same token, firms or governmental units with large discharge volumes can afford to conduct a better analysis program.

(b) Physical Properties

Temperature can be one form of stream and marine pollution that is of particular concern to a range of users--anglers, fisheries, and establishments that use cooling waters. A rise in temperature (even of 1°C) due to discharge of large volumes of used "cooling water" can have several consequences. These include fish mortality or migration, enhancement of toxic effects, as well as excessive plant growth and increased operating costs for power stations.

Transparency of the water can be decreased by pollution, with two consequences: aesthetic appearance is impaired and marine photosynthetic activity may be reduced.

Odor is perhaps the most noteworthy nuisance of polluted water. Odors due to polluted streams have been classified: putrid smell (due to hydrogen sulfide), fishy smell (probably organic amines), wormy smell (possibly phosphorus compounds), and earthy smell (due to humus). The first two odors can become very strong when streams and estuaries become deoxygenated. In some instances, addition of nitrates to streams has reduced objectionable chemical odors.

(c) Chemical Properties

pH has little value as a marine pollution parameter, though it may be a significant parameter for restricted waters. The pH of open waters may be altered within only 100 feet of the point of discharge of alkaline or acid effluents. Desirable pH limits vary with the intended use (Note 1).

Chloride Content. This is not a marine problem, but may be a factor in fresh water. There are a wide variety of standards for chloride. Some areas set limits of 8-50 mg/liter (ppm) for a drinking water; in other areas, the limit is set at 250 ppm; in still other areas, natural waters exceed any of these values. (Sea water contains about 20,000 ppm of chloride.) The tolerance of freshwater organisms to chloride content also varies: blankweed dies rapidly in a stream containing 1000 ppm of chloride; perch can tolerate up to 9000 ppm of chloride for several days, but roach can tolerate only 6000 ppm.

Nutrient Levels. As mentioned earlier (Chapter 8), pertinent nutrients include nitrogen (as ammonia, nitrite, nitrate), phosphate, and silica. Unusually great concentrations of nutrients, due to pollution, lead to high phytoplankton concentrations in estuaries or restricted waters. For example,[7] phytoplankton (but not zooplankton) concentrations increased nearly logarithmically with distance from a reference point in estuarine waters of Southern San Francisco Bay to the head water; at this point, there are several municipal waste discharges. Phytoplankton concentration averages were 317 million organisms/liter in the head waters and about 2 million/-liter at the reference point.

There is no conclusive evidence that great concentrations of nutrients are the cause of blooms of fish-killing red tides off the Florida or California coasts. These have been attributed to effects of pollution from large population areas, but blooms occur near less-populated areas and frequently occur following diatom blooms when nutrients have been depleted.[3]

Dissolved Oxygen. Significant difficulties caused by decrease in the dissolved oxygen due to waste discharge have been established for restricted waters (estuaries, bays, harbors, and streams), though not for marine waters.

Water-quality standards for dissolved oxygen (DO) represent a typical compromise.[11] The minimum DO level to prevent inhibition of fish production might be 5 mg/liter (100% optimum production). This standard could be impracticable for other uses, and a 4 mg/liter minimum would be a compromise, which would permit 75% of the optimum production.

Heavy Metals. There is a wide range of potentially toxic trace metals and much of the literature is concerned with effects on oysters and other shellfish.[6] These and other marine organisms have the ability to selectively concentrate trace metals, which has economic consequences. For example,[10] oysters tend to concentrate copper, with an enrichment factor greater than 100. Living in an environment with a relatively large copper concentration (0.13 ppm) will not kill oysters, but it does turn them an unappetizing green in about three weeks.

There is room for disagreement on the possible harmful effects of copper and other metals on marine organisms. There is no doubt that fantastic quantities of copper and other expensive metals are being discharged to the sewers each year in every industrial country.

Fluoride. The toxic characteristics of fluorides are not commonly appreciated.[12] The presence of 1 ppm (as F^-) in drinking water is necessary to prevent decay of teeth in children, but amounts greater than 1.5 ppm may have a deleterious effect on the teeth. Many natural waters are naturally polluted with fluorides in the sense that the concentration is greater than the optimum 1 ppm.

Organic Compounds. An unbelievable range of trace organic contaminants can be found in water. The man-made sources include domestic and industrial wastes and chemicals applied to the land and water. The effects of the trace organic contaminants include objectionable taste and odor, interference in water treatment (e.g., increased chlorination is required), adverse effects on aquatic life (off-taste in flesh, toxicity to fish, Note 2), and possible long-term effects on humans.

Two specific classes of organic contaminants that have proven toxic or dangerous include: oil, a fairly common and troublesome marine contaminant,[9] which tends to accumulate particles on the surface and drift to shores; and phenols, a constituent of coal-tar and many chemical wastes, which may prevent self-purification of water by destroying bacteria in restricted waters or may cause mortality of aquatic life.

24.3 Water Quality Goals

We can define environmental pollution as environmental

alteration that is detrimental to indigenous life. The definition is, admittedly, a broad one, but it does indicate the magnitude of the problem that centers on the goal of water quality.

Water quality standards and goals are often a matter of judgment that very with the locale and the particular sets of uses and problems peculiar to that locale. Often, a series of choices or "objective sets" must be advanced. For example,[11] the Delaware River Basin Commission offered five objective sets ranging from Set V (present conditions) to Set I (most stringent goals). Some are listed in Table 24-1 as an example.

TABLE 24-1

Water Quality Goals for Delaware Estuary[a]

Quantity	Set I	Set V
pH (pH units)	6.5-8.5	5.2-8.2
Alkalinity	20-120	4-51
Chlorides[b]	50-250	50-2400
Dissolved oxygen[c]	4.5-7.5	0.9-10.0
Hardness[d]	95-150	83-467
Phenols[d]	0.001-0.01	0.01-0.06

[a]Concentration unit, ppm.

[b]Maximum 15-day mean.

[c]Summer average.

[d]Monthly mean.

A further dilemma comes from conflicting uses of the

coastal zone. It is presently used as a sewage disposal system and our current instincts may be to instantly propose clearing these waters, for example, through tertiary treatment plants. At the same time, we should remember that this may have a direct, undesirable effect on fisheries. For example, to some extent, the high yields of shellfish in many estuaries (Chesapeake Bay, Long Island Sound) is due to enrichment by human waste products.[1]

Finally, a common dilemma arises with regard to the alterations that occur upon dredging channels and filling tidal flats. These may benefit certain persons, but the effects are hardly minimal and seldom beneficial. For example, in San Francisco, dredging and filling actions have considerable complications when indigenous stenotolerant species are at a disadvantage to eurytolerant ones.

24.4 Experimental Procedures

Acidity, pH. See Chapter 3.

Alkalinity. One criterion of water quality (Note 3) suggests that the total carbonate alkalinity (Chapter 4) should not exceed 120 mg/liter and that the total alkalinity, for treated water, should not exceed the hardness by more than 35 mg/liter, expressed as $CaCO_3$.

Other standards may apply for other uses, but the pertinent determinations are given in Chapter 4.

Arsenic. See Chapter 27.

Boron. Boron species are not usually considered

pollutants, but organic boron complexes evidently can interfere in determinations of alkalinity by indicator procedures. Accordingly, an independent determination of boron may be desirable. (See Chapter 31.)

Chloride Content. The chloride content can be used to classify water. The chloride content for fresh water is less than 100 ppm of $C\ell^-$ (100 mg $C\ell^-$/liter); for brackish water, 100-1000 ppm $C\ell^-$; and for salt water, greater than 1000 ppm.

Use the Harvey method (Chapter 6). The silver nitrate solution is 0.1605 M (27.26 g $AgNO_3$/liter). Use a 100-mℓ sample and titrate as usual. Calculate the chloride content:

$$\text{chloride content (ppm } C\ell^-)$$
$$= \frac{\text{m}\ell\ AgNO_3\ \text{solution} \times 0.1605 \times 35.45 \times 1000}{100\ \text{m}\ell\ \text{sample}} \qquad (24\text{-}1)$$

Dissolved Oxygen. Two determinations are of interest: Dissolved oxygen in polluted waters (Chapter 25) and biochemical oxygen demand (Chapter 26).

Fluoride. See Chapter 30.

Hardness. There are, in effect, three definitions of water hardness: theoretical, the sum of the concentrations of all metal ions (excluding those of the alkali metals) expressed as the equivalent concentration of $CaCO_3$; soap-consuming capacity, which measures the consumption of standard soap solution in terms of equivalent $CaCO_3$ concentration; and specific-cation hardness, the concentration of a specific cation, e.g., magnesium, expressed as the equivalent $CaCO_3$ concentration. The first two definitions lead to identical results, unless there are sizable concentrations of iron, aluminum, manganese, or other metal cations present.

Determine calcium hardness using EGTA (Chapter 21). Report the hardness as "hardness (calcium)" or "hardness (compleximetric, EGTA)." The appropriate conversion is: $CaCO_3$ equivalent (mg/liter) = Ca^{2+} (mg/liter) x 2.497.

Heavy Metals-Micronutrients. For drinking water, the combined manganese and iron concentrations should not exceed 0.3 mg/liter; copper should not exceed 3.0 mg/liter in acceptable water supplies (Note 3). Pertinent determinations are given in Chapters 18, 19, and 17, respectively.

Hydrogen Sulfide. (See Chapter 28).

Light Transmission. The hydrophotometer is a device for measuring photosynthetic activity and for measuring light transmitting properties of sea water. The Secchi disk is perhaps more useful in pollution monitoring programs because of its simplicity and rapidity, and because it gives results that are comparable with what humans see.[5] The Secchi disk is a round disk (diameter, 30 cm; thickness, 2 cm) which is made of steel (or wood, if weighted). The disk is painted white (Note 4). A line (100 ft, marked in 5-ft intervals) is attached to the center of the disk. The disk is lowered in the water and the depth is noted at which the disk disappears. The average of three attempts is recorded.

Nutrient Levels. See Chapters 12-14 for ammonia, nitrite, and nitrate, Chapter 10 for inorganic phosphate, and Chapter 9 for silicon. Many workers who are concerned with pollution are familiar with reporting silicon as SiO_2, nitrate as N, and phosphate as PO_4. Appropriate conversions are listed in the Appendix.

Mercury. See Chapter 32.

Organic Contaminants-Humic Acids. See Chapter 34.

Organic Contaminants-Phenols.[12] Phenolic materials are a collection of substances with hydroxy groups attached to a benzene or related aromatic nucleus which react with 4-aminoantipyrine to form an antipyrine dye. The absorbance of the resulting solution is a measure of the original phenol concentration.

Procedure[12]

(a) Reagents

Phosphoric Acid (10%). Dissolve 10 mℓ of phosphoric acid (85%) in water and dilute to 100 mℓ.

Copper Sulfate Solution. Dissolve 50 g of $CuSO_4 \cdot 5H_2O$ in distilled water and dilute to 500 mℓ.

4-Aminoantipyrine (4-AAP). Dissolve 1.0 g of reagent in distilled water and dilute to 50 mℓ. Do not store more than one week.

Potassium Ferricyanide. Dissolve 4 g of $K_3Fe(CN)_6$ in distilled water and dilute to 50 mℓ. Filter. Store in the refrigerator. Prepare a fresh solution when darkening of the solution occurs.

pH Buffer. See Chapter 21, Buffer I.

Standard Phenol Solution. Dissolve 1 g of phenol in a liter of distilled water and mix thoroughly. This solution must be standardized (see next section). The concentration of this solution is x grams/liter.

Solution A. Dilute 5 mℓ of standard phenol solution to a liter. This solution contains exactly 5x micrograms of phenol per milliliter or 5x milligrams per liter.

Bromide-Bromate Solution. Use this solution for standardization of the phenol solution. Dissolve 1.392 g of $KBrO_3$ and 5 g of potassium bromide in water and dilute to 500 mℓ.

Other Reagents. Standard thiosulfate and starch solutions are described in Chapter 7.

(b) Standardization of Phenol Solution

Place 50.0 mℓ of standard phenol solution in a 500-mℓ glass-stoppered Erlenmeyer flask, and add 100 mℓ of water. Pipet 10.0 mℓ of bromide-bromate solution in the flask, then add about 5 mℓ of concentrated hydrochloric acid (37%; sp. gr., 1.19). Stopper the flask and swirl gently. The brown color of bromine should not persist. Add additional 10.0-mℓ portions of bromide-bromate solution until the color does persist (Note 5). Stopper the flask, and after 10 minutes, add 1 g of potassium iodide.

Prepare a blank from 150 mℓ of distilled water and 10.0 mℓ of the bromide-bromate solution. Titrate the blank, then the sample with standard thiosulfate solution to a starch endpoint (see Chapter 7).

Calculate the concentration of the standard phenol solution:

$$\frac{\text{mg phenol}}{\text{liter of sample}} = (AB - C)(F)\,\frac{(1000)}{50} \qquad (24\text{-}2)$$

(AB − C) = Amount of brominating agent consumed by sample (expressed as milliliters of standard thiosulfate solution)

A = Milliliters of standard thiosulfate solution for each 10-mℓ portion of bromide-bromate reagent used for the blank.

B = Number of 10-mℓ portions of bromide-bromate used for the blank

C = Milliliters of standard thiosulfate solution used to titrate the excess reagent in the sample

F = Factor, 15.685 $\underline{N}_2$, where $\underline{N}_2$ is the actual normality of the standard thiosulfate solution (Note 6)

(c) Preliminary Purification of the Sample

The sample should be analyzed within four hours after collection; it may be preserved for a day by adding copper sulfate (1 g/liter).

Add 1 mℓ of 10% phosphoric acid solution to a 250-mℓ sample in a beaker. The pH should be 4 or less (as indicated by pH paper). Add 1 mℓ of 10% copper sulfate solution. Place the sample in an all-glass distillation apparatus (Note 7). Collect 200 mℓ of distillate. Stop the distillation and add 50 mℓ of water to the distilling flask and continue until 250 mℓ of distillate has been collected.

(d) Analysis (Note 8)

Add 100 mℓ of distillate to an Erlenmeyer flask (labeled

"sample"). Add duplicate 100-mℓ portions of phenol Solution A in Erlenmeyer flasks (labeled "standards"). Finally, add 100 mℓ of distilled water to a fourth flask (labeled "blank").

To each flask, add, successively, 2 mℓ of pH 10 buffer, 2 mℓ of 4-AAP reagent, and 2 mℓ of potassium ferricyanide reagent. Mix thoroughly after each addition.

Measure the absorbance of the standards and sample (blank in reference cell) at 510 mμ, using a 1.0-cm cell.

Calculate the phenol content:

$$\text{mg phenol/liter} = (A_x/A_s)c_s \tag{24-3}$$

where A_x and A_s are the absorbance of the sample and standard, respectively, and c_s is the concentration of Solution A (= 5x milligrams/liter).

(e) Analysis-Extraction

Prepare a working standard solution by diluting 5 mℓ of Solution A to 500 mℓ with distilled water. Transfer the working standard to a one-liter separatory funnel.

Prepare a blank by measuring 500 mℓ of distilled water into a one-liter separatory funnel. Dilute 5 mℓ of distillate to 500 mℓ and transfer to a one-liter separatory funnel.

To the standard, blank, and sample in the separatory funnels add, successively, with thorough mixing after each addition: 3 mℓ of buffer, 3 mℓ of 4-AAP solution, and 3 mℓ of potassium ferricyanide solution.

Extract the contents of each funnel with a 15-, 10-, and 5-mℓ portion of chloroform. Collect the three chloroform extracts by filtering through filter paper into a 250-mℓ graduated cylinder. Dilute to 25 mℓ with chloroform.

Measure the absorbance of sample and standard (with blank extract in the reference cell) at 460 mμ, using a 5-cm cell.

Calculate the phenol content:

$$\text{mg phenol/liter} = A_x/A_s)c_s \cdot 100 \qquad (24\text{-}4)$$

where c_s is equal to 0.05x milligrams of phenol per liter and the other terms are the same as defined before, Eqn. (24-3).

Polyphosphate. According to one view, polyphosphate concentration can be used as an index of pollution in estuaries and harbors, though it is doubtful whether it is applicable to open-ocean samples. Details are given in Chapter 29.

NOTES

1. For example, the California State Water Pollution Control Board suggested[8] the following pH limits: water contact sports, 6.8-7.2; shellfish areas, 6.8-7.2; food processing and wildlife propagation, 6.5-8.5; and industrial uses, 4.0-10.0.
2. The nondegradable synthetic detergents (syndets) once caused foaming water (including that from faucets). In some instances, waters containing syndets removed oil coating from ducks, who became waterlogged and drowned.
3. This according to the U.S. Public Health Service Drinking Water standards of 1946.

4. Some workers prefer a disk that is divided into four equal segments which are painted alternately black and white. One Secchi disk (available from G. M. Manufacturers and Instrument Corp., New York) is painted white on one side, black on the other; the black side is used in turbid waters.

5. Four 10-mℓ portions should be required.

6. The equivalent weight of phenol is equal to the molecular weight of phenol divided by the atoms of bromine per molecule of phenol (= 94.11/6 = 15.685 g) (cf. Eqn. 24-6):

$$3KBrO_3 + 5KBr + 6HC\ell \rightarrow 3Br_2 + 6KC\ell + 3H_2O \qquad (24\text{-}5)$$

$$3Br_2 + C_6H_5OH \rightarrow 2,4,6\text{-}Br_3C_6H_2OH + 3HBr \qquad (24\text{-}6)$$

Each milliliter of 0.100 N bromide-bromate is equal to 0.1 milliequivalent weight (meq) of phenol, or 15.685 x 10^{-3}g, or 15.685 mg. Each milliliter of 0.1 N bromide-bromate requires 0.1/N_2 mℓ of N_2 thiosulfate.

7. Collect the distillate in an Erlenmeyer flask marked at 200 and 250 mℓ.

8. This procedure is suitable for phenol concentrations in the range 0.1-1.0 mg/liter. The sensitivity may be increased by using the chloroform extraction procedure.

REFERENCES

1. J. H. Ryther, Environ. Letters, 1, 79 (1971)

2. G. H. Lauff (ed.), Estuaries, AAAS, Washington, D. C., 1967.

3. Advances in Water Pollution Research (E. A. Pearson, ed.), Vol. 3, Macmillan, New York, 1964.

4. C. H. J. Hull, Ibid., pp. 347-403.

5. H. F. Ludwig, and B. Onodera, Ibid., pp. 37-56.

6. J. E. G. Raymont, and J. S. Shields, Ibid., pp. 275-290.

7. E. A. Pearson, cited in ref. 5.

8. Water Quality Criteria, Calif. State Water Pollution Control Bd. Pub. No. 3, Calif. Inst. of Tech. (1952).

9. E. Føyn, Oceanogr. Mar. Biol. Ann. Rev., 3, 95 (1965).

10. P. S. Galtsoff, J. Wash. Acad. Sci., 22, 246-257 (1932), cited in ref. 6.

11. Chem. Engn. News, Oct. 17, 1966, pp. 54-60.

12. Standard Methods for the Examination of Water and Wastewater, 11th ed., American Public Health Association, Inc., New York, 1960.

25

DISSOLVED OXYGEN IN POLLUTED WATERS

25.1 Introduction

As noted in Chapter 7, modifications of the Winkler titration are necessary to determine accurately the dissolved oxygen content of polluted waters, which contain strong oxidizing or reducing agents or substantial quantities of organic materials. At least seven modifications are available which are designed to overcome problems due to specific interferences.[1] Two of the most commonly used procedures are described here.

The Alsterberg modification[2] uses sodium azide to overcome the complication caused by nitrite ion, perhaps the most common interference in polluted waters. This modification is used when the nitrite-nitrogen concentration is greater than 0.1 mg/liter and when the ferrous-iron concentration is less than 1 mg/liter; ferric-iron concentrations in the range of 200-300 mg/liter are tolerable.

The Pomeroy-Kirschman modification[3] uses a nearly saturated solution of alkaline iodide as well as sodium iodide to overcome the complications caused by high concentrations of organic substances. The modification is used when the dissolved oxygen content exceeds more than 15 mg/liter (supersaturated solution) or with domestic sewage when the water has a high organic content.

These two modifications plus a third are presented here. The last method (Section 25.4) has the advantage that further calculations are not needed.

25.2 The Alsterberg Modification

Procedure

(a) Reagents

The six reagents of the unmodified Winkler method (Note 1) are needed, as well as two other reagents.

Alkaline-Iodide-Azide Solution. Dissolve 5 g of sodium azide (NaN_3) in 20 mℓ of distilled water, and slowly add this solution with stirring to 475 mℓ of alkaline-iodide solution (Note 1).

Potassium Fluoride Solution. Dissolve 20 g of potassium fluoride dihydrate ($KF \cdot 2H_2O$) in distilled water and dilute to 100 mℓ.

(b) Procedure Modifications

There are three modifications to the procedure given in Chapter 7.

1. Use 1 mℓ of alkaline-iodide-azide reagent instead of the alkaline-iodide solution of the unmodified Winkler method.

2. Add 1 mℓ of fluoride solution just before adding concentrated sulfuric acid.

3. Use 2 mℓ of concentrated sulfuric acid instead of 1 mℓ.

25.3 The Pomeroy-Kirschman Modification

Procedure

(a) Reagents

The six reagents of the unmodified Winkler method are needed in addition to the following:

Alkaline-Iodide-Azide Solution. Dissolve 5 g of sodium azide in 20 mℓ of distilled water. Add this solution slowly with constant stirring to a special alkaline iodide solution, which is prepared as follows: Dissolve 200 g of sodium hydroxide in 250 mℓ of boiled and cooled distilled water, cool the solution to room temperature, then dissolve 450 g of sodium iodide in the alkaline solution. After the sodium azide solution has been added, the solution should be diluted if necessary to 500 mℓ (Note 2).

(b) Procedure Modifications

The sample is collected in the usual way. Add 2 mℓ of manganous sulfate solution, then 2 mℓ of alkaline-iodide-azide solution. Stopper and mix the solution. When acidifying use 1.5 mℓ of concentrated sulfuric acid (Note 3).

25.4 Direct Reading Modification

Basically, this modification ("entire bottle technique") is a variation of the Alsterberg procedure that consists in using a special concentration of thiosulfate (0.038 N) and titrating the entire contents of a standard BOD bottle (mean volume 300 mℓ).[4] On this basis, the following convenient relationship applies:

$$1\ m\ell\ 0.0380\ \underline{N}\ Na_2S_2O_3 = 1\ mg\ DO/liter \qquad (25\text{-}1)$$

The procedure has been used with an automatic potentiometric titration unit which is equipped with a glass electrode and a platinum inlay electrode.[4]

<u>Procedure</u>

(a) <u>Reagents</u>

<u>Alkaline-Iodide-Azide Solution</u>. See Section 25.2.

<u>Sulfuric Acid</u>. See Section 7.2.

<u>Starch Solution</u>. See Section 7.2.

<u>Potassium Biniodate Standard Stock</u> (0.1520 <u>N</u>). Dissolve 4.9384 g of $KH(IO_3)_2$ in 800 mℓ distilled water and dilute to one liter.

<u>Working Standard</u> (0.038 <u>N</u>). Dilute 250 mℓ of the stock solution to exactly one liter.

<u>Sodium Thiosulfate Solution</u> (stock, 0.152 <u>N</u>). Dissolve 37.73 g of $Na_2S_2O_3 \cdot 5H_2O$ in freshly boiled distilled water, dilute to about 900 mℓ, and add pellets of sodium hydroxide. Allow the solution to "age" for about 24 hours, then standardize the solution against 0.152 <u>N</u> potassium biniodate (see Section 7.2). <u>Working standard solution</u>. If the solution is exactly 0.152 <u>N</u>, dilute 250 mℓ of the thiosulfate stock to one liter. Otherwise, dilute n milliliters to one liter:

$$n\ \text{milliliters of stock standard} = 38.0/N \qquad (25\text{-}2)$$

(b) <u>Analysis</u>

Treat a sample in a 300-mℓ BOD bottle with 2 mℓ of manganous sulfate solution and 2 mℓ of alkaline-iodide-azide reagent, adding both well below the surface. Use the usual precautions to exclude air bubbles (Section 7.2). Mix well. Mix again if the precipitate settles to the bottom in less than 20 minutes. After the top one-third of the bottle is clear, carefully remove the stopper and add 2 mℓ of conc. H_2SO_4. Restopper and mix well. Titrate the entire contents of the acidified BOD bottle using 0.038 N thiosulfate solution.

(c) Calculations

The dissolved oxygen content is obtained directly from the volume of thiosulfate used, Eqn. (25-1).

NOTES

1. The directions are given in the reagents section of Chapter 7.
2. The final volume may be slightly over 500 mℓ, owing to the high concentration of salts.
3. A modification of the calculation must be made. Because of the extra volume of reagents added before the analysis, the volume of water actually used is 98.4 mℓ. For this sample, the factor is 56.71, not 56.45.

REFERENCES

1. Standard Methods for the Examination of Water and Waste-water, 11th ed., American Public Health Association, Inc., New York, 1960.
2. G. Alsterberg, Biochem. A., 159, 36 (1925).
3. R. Pomeroy and H. D. Kirschman, Anal. Chem., 17, 715 (1945).

4. FWPCA Methods for Chemical Analysis of Water and Wastes, Analytical Quality Control Laboratory, Cincinnati, Ohio, November 1969, p. 55.

26

BIOCHEMICAL OXYGEN DEMAND

26.1 Introduction

Oxidizable wastes, found in polluted waters, may be characterized by the amount of dissolved oxygen consumed under standard conditions during the aerobic biochemical action on decomposable organic matter. This amount of dissolved oxygen (expressed as milligrams/liter) is called the biochemical oxygen demand (BOD). The procedure used to determine BOD depends upon the nature and extent of pollution. Three methods are commonly used.[1]

The Direct Method. With relatively unpolluted waters (i.e., those having a BOD value of less than 8 mg/liter), the BOD is determined by measuring the dissolved oxygen before and after a standard incubation period of five days at 20°.

The Unseeded Dilution Method. With polluted waters having a BOD value of greater than 8 mg/liter, suitable portions of sample are diluted with water saturated with oxygen. The dissolved oxygen content is determined immediately after dilution and after incubation.

The Seeded Dilution Method. With sterile polluted waters which contain bactericidal substances, the effect of these substances must be neutralized and the dilution water must be seeded with the proper kind and number of organisms before true

BOD values can be obtained.

Because waters that contain sulfide, sulfite, or ferrous ions create an immediate demand on the dissolved oxygen, it is necessary to distinguish this immediate dissolved oxygen demand (IDOD) from the true BOD. The IDOD is taken to be the amount of dissolved oxygen consumed in 15 minutes in a standard water dilution. Thus, the five-day BOD is equal to the total oxygen demand minus the IDOD.

26.2 Pretreatment Considerations

The selection of the method for evaluating biochemical oxygen demand is governed by the nature of the sample. The following comments should be a useful guide.

The sample contains residual chlorine compounds. Chlorine residuals in the sample may dissipate after 1-2 hours of standing; then, the dilution method may be used. If chlorine residuals do not dissipate on standing, the neutralized sample must be treated with 0.025 *N* sodium sulfite solution (Note 1).

The sample is supersaturated with oxygen. This situation may occur during winter months or during an algal bloom. Loss of oxygen will occur during incubation, unless the dissolved oxygen content is reduced to saturation (9.17 mg/liter at 20°). This may be done by vigorously shaking a bottle partially filled with sample.

The material used for seed or the dilution water contains toxic substances. This is checked by using a standard organic compound as a test substrate. A standard glucose solution

(300 mg of glucose/liter) will have a BOD value of 224 ± 11 mg/liter; a standard glutamic acid solution (300 mg of acid/liter) will have a BOD value of 217 ± 10 mg/liter. Pipet 5.0 or 10.0 mℓ of the standard solutions into the BOD bottles, fill with dilution water, seed material, or sample, and determine the BOD, as indicated in the following procedures.

26.3 Direct Method

Procedure

(a) Apparatus

BOD Bottles. Use standard 300-mℓ bottles with ground-glass stoppers. These contain a water seal which prevents the drawing of air into the sample during the incubation period. With non-standard bottles, an effective water seal is maintained by inverting the bottle in a water bath. The bottles should be cleaned with chromic acid-sulfuric acid cleaning mixture, carefully rinsed, and allowed to drain before being used.

Constant Temperature Bath. A thermostatically controlled air or water bath is needed to maintain a constant temperature of 20 ± 1°. Light must be excluded; otherwise, dissolved oxygen may be produced by algae in the sample.

(b) Treatment of the Sample

Fill three BOD bottles with the water sample. Be sure that no air bubbles are entrapped and that the bottles are filled to overflowing when the stoppers are inserted.

Determine the dissolved oxygen concentration in one of the

three bottles, using an appropriate Winkler modification (cf. Section 25.4). Record the value obtained as the "initial DO."

The value of the five-day BOD value is the "initial DO" minus the "final DO" and should be expressed as milligrams of oxygen/liter (Note 2).

26.4 Unseeded Dilution Method

Procedure

(a) Reagents

Buffer Solution. Dissolve potassium dihydrogen phosphate (KH_2PO_4, 8.5 g), potassium hydrogen phosphate (K_2HPO_4, 21.75 g), disodium hydrogen phosphate ($Na_2HPO_4 \cdot 7H_2O$, 33.4 g), and ammonium chloride ($NH_4C\ell$, 1.7 g) in distilled water and dilute to one liter.

Magnesium Sulfate Solution. Dissolve 22.5 g of $MgSO_4 \cdot 7H_2O$ in distilled water and dilute to one liter.

Calcium Chloride Solution. Dissolve 27.5 g of anhydrous calcium chloride in distilled water and dilute to one liter.

Ferric Chloride Solution. Dissolve 0.25 g of $FeC\ell_3 \cdot 6H_2O$ in a liter of distilled water.

Dilution Water. The distilled water used in the above solutions and in preparing dilution water must be very pure, free of acid, alkaline substances, and chlorine, and must contain less than 0.01 mg copper per liter. Usually, redistillation in an all-glass apparatus will suffice.

Prepare the dilution water by adding 1 mℓ of each of the above four reagents to a liter of distilled water. Store the dilution water at 20°.

(b) Analysis Procedure

Conduct the pretreatment as necessary (Section 26.2).

Calculate the extent of dilution needed. Estimate the BOD, knowing the extent of pollution. Refer to Table 26-1.

TABLE 26-1

Data for Estimation of Dilution[1]

Type of waste	Five-day BOD range	Dilution scale	Milliliters of sample in a 300-mℓ BOD bottle
Strong industrial waste	500-5000	1:100-1:1000	3-0.3
Normal and settled sewage	100-500	1:20-1:100	15-3
Treated effluent	20-100	1:5-1:20	60-15
Moderately polluted water	10-20	1:1-1:5	300-60

A range of dilutions is desirable, and a dilution in the range of 40-90% of the original dissolved oxygen content will give the best results. At least three dilutions should be prepared, and these should be run in duplicate.

Using a pipet, accurately measure the required amount of sample into the BOD bottles. Fill the bottles completely with dilution water.

Use the appropriate Winkler modification (cf. Section 25.4) and determine the dissolved oxygen content of the diluted sample and then of the dilution water. Calculate the "initial

DO." If the volume of waste is 3 mℓ or less, the calculation is based on the dissolved oxygen content of the dilution water.

Incubate the samples at 20° for five days in complete darkness.

After five days, determine the DO of the samples. Average the dissolved oxygen concentration of the duplicate samples and record as "final DO."

Calculate the five-day BOD (mg/liter) using the equation

$$\text{five-day BOD (mg/liter)} = (\text{initial DO} - \text{final DO}) \times \frac{300\ \text{m}\ell}{\text{m}\ell\ \text{of waste per bottle}} \quad (26\text{-}1)$$

26.5 Seeded Dilution Method

Procedure

(a) Reagents

The reagents used in the preceding method are used, in addition to the following:

Seeded Dilution Water. The seed may be stale domestic sewage or lake, river, or estuary water polluted with the same waste that is under test. Typically, 1-10 mℓ of settled domestic sewage, allowed to stand 24-36 hours, is added to each liter of dilution water. With river water, 10-50 mℓ is added per liter of dilution water. Enough seeding material is added to give oxygen depletions of at least 0.6 mg/liter during the five-day incubation period. Prepare the seeded dilution water on the day it is needed. The dilution water is saturated with oxygen by aerating with clean compressed air.

(b) Analysis

The procedure used in the unseeded dilution method is followed, but an additional step is necessary: it is necessary to correct for the effect of the seed depletion of DO:

$$\text{seed correction} = \text{BOD of seed control} \times \frac{\text{\% seed in sample}}{\text{\% seed in control}} \qquad (26\text{-}2)$$

It is necessary to run a control using seed itself and following the procedure outlined for the direct method or for the unseeded dilution method, depending upon the strength of the seed. Using the seeded dilution water as a blank is unsatisfactory because of erratic results.

(c) BOD Calculation

$$\text{BOD of waste} = (\text{initial DO} - \text{final DO} - \text{seed correction}) \times \frac{300\ \text{m}\ell}{\text{m}\ell\ \text{of waste per bottle}} \qquad (26\text{-}3)$$

26.6 Immediate Dissolved Oxygen Demand (IDOD)

(a) Analysis

Prepare an additional set of duplicate (immediate demand) bottles for each of the dilutions of the standard BOD procedure. Determine the value of "initial DO" as before.

Incubate the "immediate demand" bottles for 15 minutes. Determine the value of "final DO" for these bottles after the 15-minute period. Calculate the 15-minute depletion (which is equal to the "initial DO" minus "final DO." Record the value as "IDOD."

Incubate the other bottles for five days and determine "final DO."

(b) Calculations

"Total demand" is equal to the "initial DO" minus "final DO":

$$\text{Total demand} = \text{initial DO} - \text{final DO (after five days)} \qquad (26\text{-}4)$$

The five-day BOD is calculated using the relation:

$$\text{five-day BOD} = \text{total demand} - \text{IDOD} \qquad (26\text{-}5)$$

NOTES

1. Sodium sulfite solution (0.025 N: 1.575 g of anhydrous sodium sulfite per liter of distilled water) is not stable and should be prepared as needed. To determine the amount of solution needed to neutralize residual chlorine in a 100-mℓ sample, take a 100-mℓ sample, add 10 mℓ of 10% KI solution, and titrate with sodium sulfite solution, using starch indicator (Chapter 7).

2. Allow two bottles to incubate at a constant temperature of 20° in darkness for five days. Determine the dissolved oxygen concentration of each of the bottles, average the results, and record the value obtained as the "final DO."

REFERENCE

1. Standard Methods for the Examination of Water and Wastewater, 11th ed., American Public Health Association, Inc., New York, 1960, pp. 307-324.

27

ARSENIC

27.1 Introduction

As mentioned earlier (Section 10.1), arsenic (as arsenate ion) can interfere with the analysis of inorganic phosphate. The extent of the interference is uncertain because relatively few arsenate determinations have been made concurrently with inorganic phosphate determinations for natural waters. Observations of Chamberlain and Shapiro,[1] among others, indicate the interference is serious (Note 1). Largely, the limitation has been due to inadequate methodology.

Methods are now available that permit determinations of arsenate alone[2] or simultaneously with inorganic phosphate.[3] Arsenate is reduced to arsenite by thiosulfate under acidic conditions, and the formation of colloidal sulfur is prevented by the presence of metabisulfite in large excess.

27.2 Arsenate Method

The procedure used here is a modification of one developed by Johnson[3] that permits simultaneous determination of arsenate and phosphate. Arsenate concentration is obtained as the difference between an unreduced and a reduced sample.

Procedure

(a) Reagents

Phosphate Reagents. See Section 10.2.

Sodium Metabisulfite (14% w/w, 0.86 M). Dissolve 14 g of $Na_2S_2O_5$ in 86 mℓ of distilled water. Prepare fresh each day.

Sodium Thiosulfate (0.056 M). Dissolve 14 g of $Na_2S_2O_3 \cdot 5H_2O$ in distilled water and dilute to one liter. Store in a dark bottle.

Sulfuric Acid (3.5 N). Cautiously pour 97 mℓ of conc. H_2SO_4 (96%, sp. gr. 1.84) into 903 mℓ of distilled water.

Reducing Solution. Slowly and cautiously add 20 mℓ of 3.5 N sulfuric acid solution to 40 mℓ of sodium metabisulfite solution (Note 2). Then, add, with stirring, 40 mℓ of sodium thiosulfate solution. The refrigerated solution is stable for 24 hours.

Arsenic Standard (2.5 μg-at AsO_4-As/mℓ). Dissolve 0.4500 g of KH_2AsO_4 in distilled water, and dilute to one liter.

(b) Analysis

Preliminary considerations include the following. Filter (0.45-μ membrane filter) any water samples that are visibly turbid. Prepare a set of reagent blanks using deionized water for each set of samples. Prepare enough mixing cylinders for the run (Note 3).

The following steps are involved in the treatment of samples. Measure enough 50-mℓ samples (either four or eight) in graduated mixing cylinders. Pipet 5 mℓ of reducing reagent into each cylinder of the reduced set (not the unreduced) of cylinders, stopper, and mix. After 15 minutes, pipet 5 mℓ of mixed reagent (Section 10.2) into <u>all</u> cylinders. Thoroughly mix the contents of the cylinders immediately after adding the mixed reagent. Allow at least 90 minutes for maximum color development.

Measure the absorbance at 865 mμ of each cylinder in the reduced and unreduced sets of a given sample. Use 10.0-cm cells.

The reduced samples can also be analyzed for the phosphorus content (Section 10.2).

(c) Calibration Plot

Prepare a calibration plot, absorbance of reduced working standard samples as a function of concentration (μg-at AsO_4/liter). The absorbance values should be corrected; see below, Eqn. (27-1). From this plot, calculate the slope = m and the value of the absorbance factor F (= 1/m). The value F_{ab} is about 5.1-5.2 μg-at AsO_4-As liter^{-1} absorbance^{-1}.

(d) Calculations

All calculations are based on a sample volume of 50 mℓ and a total volume of 50 mℓ. Thus, the absorbance values of the reduced set (A_R) must be corrected using the relationship

$$A_R' = A_R \text{ (correction factor)} = A_R \times 1.091 \qquad (27\text{-}1)$$

Calculate the arsenate concentration using corrected absorbance values or

$$\mu\text{g-at } AsO_4 - \text{As/liter} = (A_u - A_{uB}) - (A_R' - A_{RB}') \times F_{As} \quad (27\text{-}2)$$

Here, A_u is the mean absorbance of the unreduced sample; A_{uB} is the absorbance of the unreduced blank; A_R' and A_{RB}' are the mean corrected absorbance values for the reduced and corresponding blank, respectively; and F is the absorbance factor.

If the reduced samples and corresponding blank were analyzed for phosphorus, calculate the phosphate content from the corrected absorbance values

$$\mu\text{g-at } PO_4\text{-P/liter} = (A_R' - A_{RB}')m \qquad (27\text{-}3)$$

where m is defined as before, Eqn. (10-1).

NOTES

1. These workers[1] compared observed phosphate concentrations with estimates from an algal bioassay procedure. Appreciable differences between the two estimates were due to arsenate interference. In certain lakes, the discrepancy is disturbing (33 vs. 2.6; 26 vs. 0; 33 vs. 12.2 μg-at PO_4-P/liter). The error is evidently much less in marine samples.[3]
2. The addition should be done slowly with good mixing to avoid excessive bubbling as sulfur dioxide is evolved.
3. Two sets of 50-mℓ mixing cylinders are needed for each sample; one set for the reduced and one for the unreduced sample. The number needed per set depends upon the arsenate level. Use quadruplicate samples for deep-sea samples; use three or two per set for near-shore or estuarine samples.

REFERENCES

1. W. Chamberlain and J. Shapiro, Limnol. Oceanogr., 14, 921 (1969).

2. J. C. von Schouwenburg and I. Walinga, Anal. Chim. Acta, 37, 271 (1967).

3. D. L. Johnson, Environ. Sci. Tech., 5, 411 (1971).

28

HYDROGEN SULFIDE

28.1 Introduction

Hydrogen sulfide or hydrosulfide ion can be found in various oxygen-poor environments.[1,2] Typically, determinations of low concentrations of sulfide-sulfur are colorimetric analyses that depend upon the quantitive oxidation of sulfur in the presence of some dye or dye precursor. Methylene blue and *p*-phenylene-diamine (Lauth's violet procedure) are commonly used. The method uses a derivative of the second type and was developed by Cline.[3] It has several advantages over several previously described procedures,[3] including a single, convenient reagent and superior reagent stability at low concentrations.

28.2 Modified Diamine Method[3]

The sample is allowed to react with the mixed diamine reagent (acidified *N*,*N*-dimethyl-*p*-phenylenediamine sulfate and ferric chloride). The product, in which sulfur is incorporated, is related to a dye called Lauth's violet. The absorbance of the solution is measured and related to a sulfide standard. Salt effects are negligible and the absorbance is not temperature-dependent, within reasonable limits. By varying reagent ratios, the method can be applied to natural waters containing 1-1000 μg-at S^{2-}-S/liter.

Procedure

(a) Reagents

Hydrochloric Acid, 50% (v/v). Slowly add 250 mℓ of concentrated HCℓ (27%, sp. gr. 1.18) to 250 mℓ of distilled water. Allow the solution to cool and store in a glass-stoppered bottle.

Mixed Diamine Solution. Four different solutions are used to cover the range of sulfide concentrations (Table 28-1).

Solution A. Dissolve 4.0 g of N,N-dimethyl-p-phenylenediamine sulfate (Note 1) and 6.0 g of ferric chloride, $FeC\ell_3 \cdot 6H_2O$, in 100 mℓ of 50% hydrochloric acid (Note 2). Other solutions and the amounts to be dissolved in 100 g of acid solution are as follows.

Solution B. Diamine, 1.6 g; ferric chloride, 2.4 g.

Solution C. Diamine, 0.4 g; ferric chloride, 0.74 g.

Solution D. Diamine, 0.1 g; ferric chloride, 0.15 g (Note 1).

Thioacetamide Solution (0.1 M). Dissolve 7.50 g in distilled water and dilute to one liter. Working solution: Dilute 10 mℓ of stock solution to one liter. Prepare as needed. 1 mℓ = 10 μmole hydrogen sulfide.

(b) Sampling Procedure and Sample Storage

Two sampling procedures have been used: a syringe and a glass-stoppered bottle. Regardless of the device, manipulation of the samples and exposure to air must be minimized because of the volatility of hydrogen sulfide.

A syringe system[3] consists of a 50-mℓ syringe (with 16-gauge stainless steel cannula, Note 3) mounted in a plastic or wooden frame with a calibrated stop position for standard volume. The cannula is inserted into a rubber septum that covers the drain-cock of the sampling bottle. If possible, the colorimetric analysis should be carried out in the syringe. Otherwise, the sample should be tranferred to a 50-mℓ serum bottle.

Glass-stoppered bottles (100 mℓ) can be used. The bottles should be completely filled and the sample drawn immediately after the oxygen-analysis sample. Samples should be stored in a cool, dark place and the analysis should be started within an hour of sampling.

(c) Analysis

Transfer the sample, if necessary, to a 50-mℓ serum bottle. Using a 5-mℓ syringe (with a 4-mℓ stop position), add the diamine reagent (see Table 28-1). Mix the solution gently and allow the color to develop during a 20-minute period. The absorbance is then determined at 670 mμ.

It may be necessary to dilute samples (Table 28-1). This should be done volumetrically after the 20-minute period.

Calculate the sulfide concentration C_s, μg-at/liter, in the sample from the appropriate expression

$$C_s = F(A - A_b) \qquad (28\text{-}1)$$

Here, A is the absorbance of the sample and A_b is that of the blank. F is the conversion factor obtained from the standard

TABLE 28-1

Reagent Concentrations and Dilution Factors for Various Sulfide Concentrations

Sulfide concentration, μmoles/liter	Mixed diamine solution	Dilution factor, mℓ:mℓ	Suggested cell, cm	Working standard, mℓ/50 mℓ
250-1000	A	1:50	1	30
40-250	B	4:50	1	10
3-40	C	25:50	1	1.5
1-3	D	25:50	10	0.2

sample.

Calculate the formal hydrogen sulfide concentration using the expression

$$\text{m}\ell\ H_2S\ \text{(STP)/liter} = C_s \times 0.0224 \qquad (28\text{-}2)$$

(d) Standardization (Note 4)

Prepare standard solutions by diluting x milliliters of working thioacetamide solution to 50 mℓ with distilled water in a stoppered mixing cylinder (see Table 28-1 for suggested values of x). Heat the cylinders in a boiling water bath for about 10 minutes to generate the hydrogen sulfide. When the solutions are cooled to room temperature, carry out the analysis in duplicate.

Calculate the value of F from

$$F = 20x/(A_s - A_b) \qquad (28\text{-}3)$$

Here, x represents the milliliters of working standard, A_s is the average absorbance of the standards, A_b is the absorbance of the blank.

NOTES

1. Eastman Kodak No. 1333. Recrystallization is not necessary. The oxidation may vary from one lot to another, though the calibration factor seems to remain constant. According to Nusbaum,[4] the oxalate salt leads to greater reproducibility.
2. Store solutions in a dark bottle in a refrigerator. Solutions B and C should be prepared as needed; the others are stable for several weeks.
3. Coat the cannula with siliclad to limit corrosion.
4. Cline,[3] among others, uses $Na_2S \cdot 9H_2O$. If improved accuracy is required, the sulfide or thioacetamide working solutions can be standardized iodometrically.[5]

REFERENCES

1. R. Pomeroy, Sewage Works J., _8_, 572 (1936).
2. F. A, Richards, Deep-Sea Res., _7_, 163 (1960).
3. J. D. Cline, Limnol. Oceanog., _14_, 454 (1969).
4. I. Nusbaum, Water Sewage Works J., _112_, 113, 150 (1965).
5. M. S. Budd and H. A. Bewick, Anal. Chem., _24_, 1536 (1952).

29

POLYPHOSPHATE

29.1 Introduction

The presence of polyphosphate may be a useful indicator of one type of pollution in certain estuaries or harbors. For example, the amounts of polyphosphate in San Diego harbor generally decreased seaward, according to Solórzano and Strickland.[1] In the more contaminated parts of the harbor, considerable amounts of polyphosphate were found that may have been due to sewage from ships.

In contrast, in samples from coastal samples, even from eutrophic waters, the observed polyphosphate content is low.[1,2] Occasionally, higher samples have been found in red-tide outbreaks.[1]

This pattern--absence in marine water, presence in polluted areas--is reasonable if it is hypothesized that, in relatively unpolluted marine environments, polyphosphate comes from phytoplankton and that any polyphosphate lost is soon reused by bacteria or other plants.

Unfortunately, few data are available for marine species, but the hypothesis is supported by a study of two plankton species (Skeletonema costatum and Amphidinium carteri). In bacteria-free cultures, both algae had a polyphosphate content that was about 3% of cellular phosphate in phosphate-rich medium, phosphate-starved cells contained a smaller fraction, and cells could remove added polyphosphate from the medium. In addition, with the second alga, orthophosphate accumulated in the medium. Presumably, the alga was able to hydrolyze external polyphosphate more rapidly than it could be utilized.

29.2 Irradiation Method for Polyphosphate[1]

Polyphosphate in sea water can be determined by irradiation and acid hydrolysis of a sample. Irradiation destroys the organic material and subsequent hydrolysis of polyphosphate yields inorganic phosphate. Irradiation will not depolymerize polyphosphate, nor will storage in sterile sea water at room temperature. At higher temperatures, produced during irradiation, slow hydrolysis will occur. Thus, the irradiation time must be limited to avoid hydrolysis.

The method consists in irradiating a sample of sea water, dividing the irradiated sample, hydrolyzing one, and determining the difference in orthophosphate content. The difference is equated to the polyphosphate concentration.

Procedure

(a) Apparatus (See Section 15.2.)

(b) Reagents

Hydrochloric Acid. Add 100 mℓ of concentrated hydrochloric acid (37%, sp. gr. 1.18) to 150 mℓ of distilled water.

Hydrogen Peroxide. Use 30% reagent-grade.

(c) Glassware

It is worth reemphasizing that no glassware or containers

should be washed with phosphate-based detergents. Glass storage bottles should be used and washed in the special way described for phosphorus analyses (Section 10.2).

(d) Analysis

The first step consists in destroying organic matter by irradiation. Treat 80 mℓ of filtered (0.45-μ membrane filter) sample with 1-2 drops of hydrogen peroxide. Irradiate the sample in a photochemical reactor. The temperature should not exceed 70°, and irradiation time should be less than one hour (Note 1). Cool the irradiation tube to room temperature. Add enough phosphate-free water to bring the volume to the initial 80 mℓ. Remove 40 mℓ for phosphate analysis (Section 10.2).

Next, hydrolyze any polyphosphate. Add 1 mℓ of hydrochloric acid to the remainder and warm in a 90° bath for two hours. Allow the tube to cool. Bring volume to the initial level.

Analyze both hydrolyzed and unhydrolyzed aliquots. Add 1 mℓ of hydrochloric acid to the nonhydrolyzed aliquot, and analyze both samples for phosphate (Section 10.2). The difference between the phosphate concentrations of the two samples corresponds to the concentration of orthophosphate.

NOTE

1. The irradiation period is based on a "189A" mercury arc used by Armstrong and co-workers[3] and different times will be needed for other units (Section 15.2). Also, larger periods of irradiation may be needed to determine higher concentrations of polyphosphate. The particular unit used should be evaluated for correct irradiation time. A medium-length polyphosphate polymer (e.g., 20 phosphorus atoms per chain, Monsanto, St. Louis, Mo.) should be used to determine that the recommended irradiation time does not produce more than 10% error.

REFERENCES

1. J. Solórzano and J. D. H. Strickland, Limnol. Oceanog., 13, 515 (1968).

2. F. A. J. Armstrong and S. Tibbits, J. Mar. Biol. Assoc. U.K., 48, 143 (1968).

3. F. A. J. Armstrong, P. M. Williams, and J. D. H. Strickland, Nature, 211, 481 (1966).

30

FLUORIDE

30.1 Introduction

There is a widespread interest in the fluorine content of natural waters. In the United States, over 2000 cities with a combined population in excess of 40 million add fluorides to water as a caries preventative (0.8-1.2 mg/liter). Naturally occurring fluorides and industrial discharges often increase the fluoride levels beyond the recommended level (1.5 mg.liter). The variation of fluoride in sea water has also been of considerable interest, though until recently, a suitably precise method for the estimation of fluoride was not available.

Many determinations[1] are based on the ability of fluoride ion to destroy the colored organic complexes of zirconium (among other metals) through the formation of a more stable fluoride complex. The decrease in color would be a measure of the amount of fluoride present. Unfortunately, the method is sensitive to the concentration of sulfate ion, which also forms complexes and thus has a bleaching action. Various methods have been devised to remove or compensate for sulfate. These include raising the level of sulfate to a constant value, precipitating the sulfate, separating the fluoride by steam distillation or by using an anion-exchange resin. All processes tend to be time-consuming and to require close attention to detail, particularly at low fluoride levels. Many of these difficulties are overcome in a colorimetric method of Greenhalgh and Riley[2] (Section 30.2).

Two noncolorimetric methods for determining fluoride have been reported. The first, neutron activation analysis, can be used advantageously when equipment is available and when fluoride must be determined in a range of samples (e.g., sea water, sea weeds, fossils, and sediments).[3] With sea water, the precision is probably less than the colorimetric method. The second, use of a fluoride-selective electrode (Section 30.3), has several advantages if typical sea water is analyzed.[4] The analyses with this electrode are rapid, and the equipment is relatively inexpensive and readily adaptable to field use. Error limits are about 5%.

30.2 <u>Colorimetric Method for Fluoride</u>

This method is based on the formation of a fluoride complex with a lanthanum-alizarin reagent. The fluoride complex is formed rapidly at a pH of 4.50, is stable for several days, and follows Beer's Law up to 2.0 μg F^-/mℓ. The method showed a standard deviation of 0.0038 μg F^-/mℓ for sea water having a mean fluoride content of 1.371 μg F^-/mℓ (11 replicate determinations).[2]

<u>Procedure</u>

(a) <u>Reagents</u>

<u>Lanthanum-Alizarin Complexone Reagent</u>. Dissolve 0.0479 g of alizarin complexone (Note 1) in 0.1 mℓ of concentrated

ammonium hydroxide (28%, sp. gr. 0.90), add 1.0 mℓ of 20% w/v ammonium acetate (Note 2), followed by 5-10 mℓ of water. Filter the solution (Note 3) either into a 200-mℓ graduated cylinder equipped with a stopper or a 200-mℓ volumetric flask. Add 8.2 g of anhydrous sodium acetate (reagent-grade) and 6.0 mℓ of glacial acetic acid. Wash the filter paper with distilled water. Use enough distilled water to dissolve all solids in the cylinder. Add 100 mℓ of acetone slowly with swirling.

Dissolve 0.0408 g of lanthanum oxide (spectrographic grade) in 2.5 mℓ of 2 M hydrochloric acid (Note 4). Gentle warming of the mixture will help.

Add the lanthanum solution to the alizarin solution, and dilute to 200 mℓ with distilled water. Mix well, and after 30 minutes, adjust the volume if necessary (Note 5).

Standard Fluoride Solution. Solution A (1 mg F^-/mℓ). Dissolve 2.210 g of anhydrous sodium fluoride in 50 mℓ of distilled water containing 1 mℓ of 0.1 M sodium hydroxide solution (Note 6); dilute to one liter.

Solution B (10 μg F^-/m). Dilute 10.0 mℓ of Solution A to one liter with distilled water. Prepare Solution B fresh each day.

Acetic Acid (6 M). Dilute 35 mℓ of glacial acetic acid to 100 mℓ with distilled water.

(b) Analysis

To 15 mℓ of filtered (Note 3) sea water contained in a

50-mℓ graduated mixing cylinder, add 8.0 mℓ of the lanthanum-complexone reagent. Add (0.4-0.02x) milliliters of 6 M acetic acid, where x is the chlorinity of the water (Note 7). Dilute to 25 mℓ with distilled water. Mix well.

Prepare a blank solution by adding 8.0 mℓ of lanthanum-complexone reagent, and 0.4 mℓ of 6 M acetic acid to 15 mℓ of distilled water in a second mixing cylinder. Dilute to 25 mℓ. After 30 minutes, measure the absorbance of the sample solution at 622 mμ (1-cm cell) with the blank solution in the reference cell (Note 8).

(c) Calibration Curve

Prepare calibration solutions as follows: Pipet 0.0, 1.0, 1.5, and 2.0 mℓ of fluoride Solution B in separate mixing cylinders. Add distilled water (Note 9) to make the volume 15 mℓ. Repeat the analysis procedure as outlined.

Determine the fluoride content of an unknown sample from the plot of absorbance versus micrograms of fluoride in a 15-mℓ sample.

(d) Application to Nonsaline Waters

The method can be applied to most nonsaline waters without difficulty. The only serious interference is the presence of abnormally large amounts of aluminum (greater than 2 ppm). In this case, the interference is due to the stability of the aluminum-fluoride complex. The aluminum can be removed by solvent extraction with oxine using a standard procedure.[2]

30.3 Fluoride Analysis with a LnF_3 Electrode[4]

The LnF_3 electrode is an example of a solid-state electrode. Apparently, it consists of a single LnF_3 crystal, doped with Eu^{2+}, in contact with internal and external solutions. (The composition of the internal solution is proprietary information.) This electrode has a useful response to at least 10^{-6} M fluoride. Hydroxide ion is the only major interference. The selectivity of the electrode for F^- over OH^- is about tenfold and the selectivity for F^- over other anions ($C\ell^-$, Br^-, I^-, HCO_3^-, NO_3^-S, SO_4^{2-}, and HPO_4^{2-}) is at least a thousandfold.

In the determination of fluoride by the LnF_3 electrode, two problems are minimized by means of a buffer. The potential interference by hydroxide ion is minimized by a pH buffer, and interference from other ions seems to be minimal. The problem of variations in ion composition is also minimized by means of a special buffer (TISAB). The electrode stability is good, and any drift of the standardization can be checked easily using the standard fluoride solution.

Procedure[4]

(a) Equipment

Electrodes. Use a lanthanum fluoride electrode (Orion 94-09) together with a standard calomel reference electrode.

Potentiometer. Use a Corning Model 12 research pH meter or equivalent.

(b) Reagents

Artificial Sea Water (see Section 4.2). Determine the

salinity (= S_1) by means of the Harvey method (Section 6.3).

Standard Fluoride Solution (10 ppm). Add 500 mℓ of artificial sea water to 0.1105 g of NaF (dried for two hours at 110°).

Total Ionic Strength Activity Buffer (TISAB). To 500 mℓ of distilled water, add 57 mℓ of glacial acetic acid, 58 g of NaCℓ, 0.30 g of sodium citrate, and stir. Add 3 M NaOH (Section 8.4) dropwise with stirring until the pH is 5-5.5, then dilute to one liter with distilled water.

(c) Analysis

First standardize the electrode system. To 100 mℓ of standard fluoride (C_1 = 10 ppm) solution ($S^0/oo-S_1$) add 20 mℓ of buffer solution (TISAB) and record the emf reading (E_1) with a fluoride-calomel electrode. Calculate the constant β for this system:

$$E_1 = \beta - 2.303(RTF^{-1}) \log_{10}C_1 \qquad (25\text{-}1)$$

As noted previously (Section 4.2), the value of RTF^{-1} at 25° is 59.16 mV, the value of 2.303 $\log_{10}C_1$ is 2.303.

Next, use a 100-mℓ seawater sample of unknown fluoride concentration C_2 and a salinity of S_2. Add 20 mℓ of TISAB buffer and again measure the cell potential (= E_2).

Calculate the value of C_2. If $S_2 = S_1$, $C_2 = C_x$ then

$$E_2 = \beta - 2.303(RTF^{-1}) \log_{10}C_x \qquad (25\text{-}2)$$

If $S_2 \neq S_1$, as is more likely, then

$$C_2 = Q_aC_x \qquad (25\text{-}3)$$

Here, Q_a is a function of salinity and

$$Q_a = (S_2/55.8) + 0.385 \tag{25-4}$$

If the value of S_2 is between 30 and 36, the value is given by Eqn. (25-4); if S_2 is between 36 and 39, the following relationship applies:

$$Q_a = (S_2/49.4) + 0.300 \tag{25-5}$$

It should be noted that the correction term Q_a becomes significant when ionic compositions of standard and sea water differ, because the ratio of fluoride activity to total concentration differs from the standard to the sample.

NOTES

1. Alizarin complexone: 1,2-dihydroxyanthraquinonyl-3-methylamine-N,N-diacetic acid or (3,4-dihydroxy-2-anthraquinonyl)-methyliminodiacetic acid, may be prepared[5] or purchased from Aldrich Chemical Co., Inc., 2371 North 30th Street, Milwaukee, Wisconsin. If purchased as alizarin complexone dihydrate (12,765-5) use 0.0524 g instead of 0.0479 g.
2. Dissolve 20 g of ammonium acetate in distilled water and dilute to 100 mℓ.
3. Filter through Whatman No. 1 paper.
4. Add 13 mℓ of concentrated (37%) acid to distilled water, dilute to 100 mℓ.
5. The solution is stable for at least a week.
6. Dissolve 0.4 g (about two small pellets) in distilled

water, dilute to 100 mℓ. Store in a plastic bottle.

7. The chlorinity need be known only within five units.[2]

8. The apparent pH of the blank and sample solutions, when diluted to volume, should be 2.50 ± 02.

9. It may be more convenient to use distilled water instead of artificial sea water of appropriate chlorinity. If so, a salt correction must be applied to the absorbance. To calculate the expected absorbance for sea water with chlorinities of 15-20°/oo, multiply the absorbance for distilled water solutions by 1.040. Other factors are 1.029, 1.012, and 1.00 for samples having chlorinities of 10.13, 5.05, and 0.00 parts per thousand, respectively. Correction factors for other chlorinities may be interpolated.

REFERENCES

1. Standard Methods for the Examination of Water and Wastewater, 11th ed., American Public Health Association, Inc., New York, 1960, pp. 122-131.

2. R. Greenhalgh and J. P. Riley, Anal. Chim. Acta, 25, 179 (1961).

3. P. K. Wilkniss and V. J. Linnenbom, Limnol. Oceanog., 13, 530 (1968).

4. T. B. Warner, Science, 165, 178 (1969).

5. R. Belcher, M. A. Leonard, and T. S. West, J. Chem. Soc., 2390 (1958).

31

BORON

31.1 Introduction

The contribution of boron species to the alkalinity of a seawater sample was considered earlier Eqn. (4-1). The contribution of borate ion can seldom be neglected, e.g., only at pH 7.3 or less. It is assumed that the boron-chlorinity ratio is constant when alkalinities are corrected for borate. The assumption may be erroneous when dealing with nearshore waters or with water taken from the oxygen minimum layer. In both cases, it appears that variations in the $B(mg/kg)/C\ell^{o}/oo$ ratio is due to the presence of unknown organic compounds that can complex boron. It is thus useful to have an independent measurement of boron.

Three chief methods have been used in the past to determine boron: a mannitol method, a fluorimetric method, and a (curcumin) colorimetric method.

Mannitol and other polyhydroxy compounds (including presumably the unknown organic compounds mentioned above) react with boric acid to form a monobasic acid that can be titrated. A reasonable representation of the reaction is

$$\begin{array}{c} -C-OH \\ | \\ -C-OH \end{array} + H_3BO_3 \rightarrow \begin{array}{c} -C-O \\ | \\ -C-O \end{array}\!\!\!\!\!> BOH + 2H_2O \qquad (31\text{-}1)$$

Most of the principal determinations of boron in sea water have used a mannitol method, most recently by a titrimetric procedure (*cf*. Hood and Noakes[1]). One such method has a coefficient of variation of about ± 1%.[1] The procedure gave low results in the presence of competing organic compounds,[2] unless the organic substances are oxidized.[1,2]

A fluorimetric determination of boron in sea water with benzoin required that the sample be pretreated with ion-exchange resin and had a coefficient of ± 2%.[3]

The curcumin method is probably the most sensitive procedure for determining trace amounts of boron. With suitable modifications and precautions, it has a dual advantage of simplicity and sensitivity. Moreover, polyhydroxy compounds do not interfere; a procedure developed by Greenhalgh and Riley has a low coefficient of variation (±0.45%).[4]

31.2 Modified Curcumin Method

The method used here is based upon modifications made by Uppström[5] and others.[6,7] A direct curcumin method was developed by Hayes and Metcalfe,[6] and several improvements were made by Grinstead and Snider,[7] including the elimination of an evaporation step in the presence of sodium hydroxide, elimination of a filtration step before absorbance measurement, and diminution of color due to excess reagent.

The various modifications have led to convenience, rapidity, and reduced sample size, and thus enhanced sensitivity. A salt error is not observed over a reasonable range of salinity and the method is applicable to about 0.05 ppm of boron.

Procedure

(a) Reagents (Note 1)

Curcumin Solution. Dissolve 0.125 g of curcumin (Eastman 1179) in 100 mℓ of glacial acetic acid.

Acid Reagent. Cautiously add 50 mℓ of concentrated sulfuric acid (96%) to an equal volume of glacial acetic acid.

Acetate Buffer. Dissolve 250 g of ammonium acetate and 300 mℓ of glacial acetic acid in distilled water and dilute to 1.1 liters.

Boron Standard. Dissolve 0.0229 g of boric acid in one liter of 0.5 M NaCℓ solution. The stock solution contains 4.0 mg of B/liter of solution.

(b) Analysis

Pipet a 0.200-mℓ sample (Note 1) into a 100-mℓ polyethylene bottle. Add 3 mℓ of curcumin solution and 3 mℓ of acid reagent. Mix the contents and allow the mixture to stand for 0.75 hour. Then, add 15 mℓ of acetate buffer, mix well, and allow the solution to stand 0.5 hour at room temperature. Measure the absorbance at 555 mμ in 1-cm cells, using a blank as reference solution (Note 2).

(c) Standardization

Prepare a calibration plot by taking n milliliters of stock solution and diluting to 40 mℓ (with 0.5 M NaCℓ) to prepare additional standard solutions. The concentration is equal to (40/n) x 4 mg B/liter. Suggested values of n are 5, 10, 15,

25, 30, and 40. The calibration plot should be linear in the range 0-4 mg B/liter.

NOTES

1. All reagents and distilled water must be stored in polyethylene bottles. It is helpful if a low-boron reagent-grade sulfuric acid is used.
2. Run parallel reference samples using 0.200 mℓ of boron-free NaCℓ solution.

REFERENCES

1. D. W. Hood and J. E. Noakes, Deep-Sea Res., 8, 121 (1961).
2. J. A. Gast and T. G. Thompson, Anal. Chem., 30, 1549 (1958).
3. W. J. Barnes and C. A. Parker, Analyst, 85, 828 (1960).
4. R. Greenhalgh and J. P. Riley, Analyst, 87, 970 (1962).
5. L. Uppström, "Analyses of boron in sea water by a modified curcumin method," in D. Dyrssen et al., Report on the Chemistry of Sea Water, IV, Department of Anal. Chem., Univ. Goteborg, Sweden, July 12, 1967.
6. M. R. Hayes and J. Metcalfe, Analyst, 87, 956 (1962).
7. R. R. Grinstead and S. Snider, Analyst, 92, 532 (1967).

32

MERCURY

32.1 Introduction[1-4]

The toxicity of mercury and mercury compounds has produced much interest in the detection and determination of this element. But how to analyze for environmental mercury is only one problem, albeit an admittedly major one. The sources of environmental mercury, where it goes, what reactions it undergoes, and what the toxic levels are to man and lower organisms, are related problems that need to be considered as well.

The effects of mercury poisoning in man can be reversible (i.e., total recovery is possible without residual effects), but can also be irreversible, and can even be fatal. Some symptoms characteristic of each level are listed in Table 32-1.

Concentrations at which various effects occur are not listed in Table 32-1 because wide variation in individual susceptibility exists. The allowable daily intake (ADI) of mercury was set at 0.4 μg per kg of body weight by the Swedish Commission on Evaluating the Toxicity of Mercury in Fish. This includes a safety factor of only 10. A 70-kg man (154 ℓb) would have an ADI of 0.03 mg, or he could eat only 60 g (2.1 oz) of fish containing the limit of 0.5 ppm each day (Note 1). Some adults may lack symptoms of poisoning at 10 or even 100 times this level, but children are much more sensitive. The various conclusions based upon mercury-intake data depend upon uptake mechanisms.

TABLE 32-1

Some Effects and Symptoms of Mercury Poisoning

Effects	Symptoms
Reversible	Fatigue, headaches, memory impairment, loss of ability to concentrate
Irreversible	Loss of motor control, e.g., hands, speech, movement; loss of sensory operations (hearing and visual)
Fatal	Death, preceded by blindness, mental and emotional deterioration, involuntary mobilization, loss of consciousness

Mercury is a cumulative poison and is only slowly removed from human systems. According to one view, it takes a year to achieve a steady-state condition (mercury intake versus excretion, assuming constant mercury-containing diet); another year would be required for 99% of the mercury to be removed.

Present data suggest that mercury is widely distributed in the oceanic food chain and is at a detrimental level. For example, the concentration in tuna and swordfish taken from oceanic waters near Malaya exceeded the 0.5 ppm mercury limit recommended by the U.S. Food and Drug Administration. Also, certain fish from the Great Lakes contained 7 ppm of mercury in 1970. Finally, preliminary results indicate mercury compounds can reduce the growth rate of phytoplankton and diatoms under laboratory conditions at very low concentrations (0.1 ppb).[2] Toxic doses are higher, of course (Table 32-2).

What conclusions validly emerge depend upon the validity of the analytical methods and comparison with past results.

TABLE 32-2

Lethal Concentrations of Mercuric Chloride, $HgCl_2$, for Various Organisms[2]

Type	Example	Lethal concentration, ppb
Bacteria	*E. Coli*	200
Phytoplankton	*Scenedesmus*	30
Zooplankton	*Daphnia magna*	20
Flatworm	*Polycelis nigra*	270
Mollusca	*Australorbis glabratus*	1000
Fish	Guppy	20
	Rainbow trout	9200
Man	Adult, chronic	one million

32.2 Mercury Sources

Mercury from various sources has become widely distributed in the environment. In general, the source of discharge varies depending on the use of mercury. Introduction of mercury occurs indirectly or directly, as well as deliberately or accidentally. Some examples will indicate the magnitude of the problem.

The use of mercury is widespread, and losses to the

environment seem to be inevitable. About 3000 metric tons (over six million pounds) were consumed in the United States in 1969 (Table 32-3) and excessive direct discharge of waste materials occurred in the past. For example, at one time, the chlor-alkali industry discharged into the environment 0.25-0.5 ℓb of mercury per ton of sodium hydroxide produced. Some discharge must also arise from widespread use of mercury as a catalyst, e.g., in the manufacture of plastics (urethane and vinyl chloride) or chemicals such as acetaldehyde. In the early 1950's, the discharge of mercury compounds by a plastics plant in Minamata Bay, Japan evidently contaminated fish and shellfish and over 50 persons were severely poisoned (Note 2).[1,5] Mercury is also released indirectly into the environment as a result of the widespread use of either elemental mercury (e.g., in thermoregulators) or its compounds. Burning mercury-containing paper is one example of indirect release. Mercury compounds find considerable use as pharmaceuticals and agricultural agents (fungicides and antibacterial agents), and these are lost to the environment. Paper and pulp mills used to use phenyl mercury acetate (C_6H_5HgOAc) as a slimicide. Though this practice ceased in 1959, they still use sodium hydroxide which contains trace amounts of mercury.

The magnitude of mercury release may be much greater than the data in Table 32-3 would indicate. The discrepancies in mercury inventories (1944-1958) are listed in Table 32-4. For example, at the end of 1944, the inventory of mercury was

80,900 flasks. The net accumulation of mercury for 1945 can be calculated from the production, import, export, and consumption values (Eqn. 32-1) as 36,000:

TABLE 32-3

Mercury Consumption in the U. S.[a]

Use	Consumption, metric tons
Electrolytic chlorine	716
Electrical apparatus	644
Paints	336
Instruments (industrial and control)	241
Dentistry	105.5
Catalysts	102
Agriculture, industrial fungicides and bactericides	93
General laboratory use	70.5
Pharmaceuticals	25
Pulp and paper-making	19
Amalgamation	7
All other uses	334.5
Total	2694

[a]U. S. Bureau of Mines data.

$$\text{net mercury accumulation} = \underset{(30{,}800)}{\text{production}} + \underset{(68{,}000)}{\text{import}} - \underset{(1{,}000)}{\text{export}} - \underset{(62{,}400)}{\text{Consumption}} \qquad (32\text{-}1)$$

TABLE 32-4

U.S. Mercury Inventory Discrepancies,[4] Flasks[a]

Year	Production	Import	Export	Consumption	Final year inventory[b]	Discrepancy
1944-1946	93,900	102,100	2,600	136,900	39,900	-83,800
1947-1949	47,500	148,000	2,000	121,700	21,000	-90,700
1950-1952	24,300	175,800	1,000	148,600	34,400	-37,100
1953-1955	51,800	168,800	1,900	152,300	10,000	-90,800
1956-1958	94,400	108,300	3,700	161,600	35.100	-12,300
Total	311,900	703,000	11,200	721,100	-45,800	-314,700

[a]Each flask contains 76 ℓb of elemental mercury.

[b]Initial year inventory, 80,900 flasks.

TABLE 32-5

Mercury in the Environment

Sample	No. of sample	Mercury range	Concentration, ng/g mean	Source[a]
Crustal				
Swedish	273	4-929	60	1
California	—	20-40	—	2
Igneous rocks, sandstone	—	30-80	—	1
Coal				
Germany	11	1.2-25	12	2
Donets Basin, USSR	206	50-10,000	1100	2
Atmospheric				
Snow, near Stockholm	—	0.08-5	—	1
Air over Pacific 20 miles offshore	—	0.6-0.7	—	2
Air, Chicago area	22	3-39	9.7	2
Rainwater	—	—	—	1
Sea water				
Atlantic, Indian, Red Sea, etc.	14	0.7-2.0	1.1	2
Pacific, Ramago Deep	4	0.15-0.27	0.2	2

[a]Ref. 1, pp. 65-70; Ref. 2, pp. 53-62.

The expected inventory for 1945 should have been

$$\text{discrepancy} = \text{observed inventory} - (\text{net accumulation} + \text{previous inventory}) \quad (32\text{-}2)$$

36,000 + 80,900 or 116,900 flasks; the observed inventory was 83,900, which means a discrepancy of -33,000 flasks. In the 15-year period 1944-1958, the discrepancy is -314,700 flasks of mercury, or 10,900 metric tons. Does this represent careless bookkeeping, or is it a measure of the mercury loss to the environment?

In view of the documented and suspected input of mercury to the environment, many have called for strict bans to be placed on mercury use. Two problems preclude strict bans from being effective. First, many industries are extremely dependent on mercury (Table 32-3) and would collapse without it, though no doubt substitutes could be found in many instances. Second, strict bans, even if feasible, would eliminate only about 50% of the input. Normal geological loss from rocks, fossil fuels, etc. contributes the other 50% of total input (Table 32-5).

32-3 Forms of Mercury

Several facts suggest that the concentration of certain species of mercury, rather than total mercury content, is the crucial factor in the mercury situation.

Different forms of mercury are found in the environment:

$$\underset{(1)}{Hg^{\circ}} \longrightarrow \underset{(2)}{\overset{(Hg^{+})}{Hg^{2+}}} \longrightarrow \underset{(3)}{H_3CHg^{+}X^{-}} \longleftarrow \underset{(4)}{C_6H_5Hg^{+}} \qquad (32\text{-}3)$$

Elemental mercury (1) can be oxidized under aerobic conditions to inorganic mercury (2) or maybe to methyl mercury ion (3), an alkyl mercury; a fourth form, aryl mercury, of which phenyl mercury ion (4) is one example, can be converted to methyl mercury. All four forms differ in their toxicities and other properties. For example, alkyl mercury and especially methyl mercury are the most toxic, much more so than phenyl mercury and inorganic mercury compounds. The damage from phenyl mercury and inorganic mercury compounds is reversible, but alkyl mercury compounds can cause irreversible damage to the central nervous system. Finally, the absorption rates from food are profoundly different: 98% of methyl mercury in food can be absorbed by tissue; the value for inorganic mercury is about 1%.

The rate of methylation of mercury and inorganic mercury in the environment is influenced by several factors, including season, pH, and biota. Methylation rates appear to be greatest in the winter and early spring, according to Swedish studies, and methylation decreases during summer and fall. This trend has been related to a seasonal ability of organisms to methylate. The methyl mercury concentration in natural water is evidently greater in waters of low pH, possibly because dimethyl mercury decomposes to methyl mercury at low pH. Suitably low pH conditions may exist in anaerobic bottom sediments of streams and lakes.

How biochemical methylation occurs is presently unknown, but it has been diligently studied[1,6,7] and reasonable mechanisms have been advanced. Obviously, a methyl transfer agent is required. Of three known methyl transfer coenzymes (Note 3), methyl corrinoids (I, Figure 32-1) seemed most likely. Kinetic studies indicated that Hg^{2+} reacts with the corrinoid in two stages, one rapid, one slow. Recent work (Note 4) suggests that the rapid reaction is nonenzymatic. Methyl transfer is pH-dependent, and a second mercuric ion slowly attacks the CH_3^- ion (Figure 32-1). Methylation does not occur unless the Hg^{2+}/methyl corrinoid ratio is 2. It also appears that the oxidation number of cobalt remains constant.

The means by which mercury is transported from sludge to fish is also by no means well understood. The inability to detect methyl mercury in the intermediate waters is an analytical limitation and contributes to the mechanistic uncertainty.

A modified mechanism is necessary to explain methylation under anaerobic conditions (Figure 32-1). Under these reducing conditions, mercurous mercury and mercuric ion are in equilibrium. Photolysis of light-sensitive methyl corrinoids produces a cobaltous species and a methyl radical that can

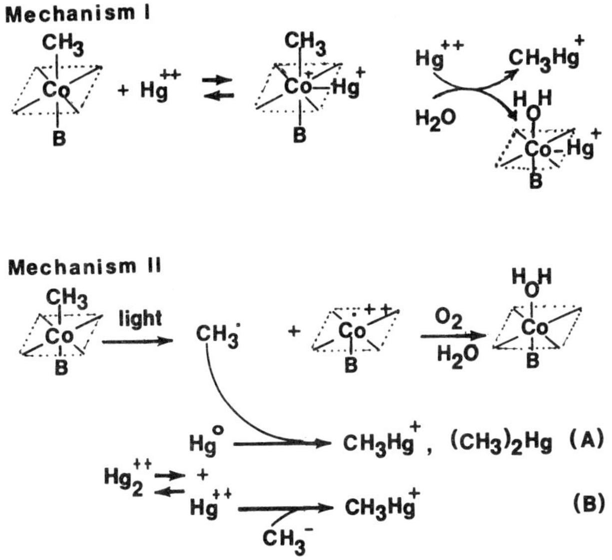

FIG. 32-1

Generalized mechanisms for methylation of mercury. Here B is a base, Mechanism II represents a photochemical reaction that has (A) an enzymatic, anaerobic pathway and (B) a chemical, aerobic pathway.[3]

react with mercury to produce methyl mercury or dimethyl mercury. It is an interesting thought that DDT may inhibit methylating enzymes.

Concentration of mercury by organisms may occur indirectly, through the food chain, or directly, through assimilation. The pathways by which mercury is transported from the sediment or from the atmosphere into fish is uncertain, though these are matters that are receiving intense study.

It is possible that some of these studies may have serious flaws, and some may raise interesting questions. Some examples may illustrate this. Methylation may not relate directly to mercury contact in sediment, possibly because of inhibitors or the absence of suitable biota. The levels of mercury in fish from some museum samples taken over 50 years ago may have high levels of mercury, possibly because mercury compounds were present in the preservative. Finally, though conversion of mercury in aquaria sediments to methyl and dimethyl mercury is a well-documented observation, this does not automatically provide evidence for the food-chain transfer to fish in the aquaria. It is reasonable to suppose that methylation can occur *in vivo* in these fish; at least, until analytical techniques permit direct analysis of the aquaria water.

32.4 Analytical Methods

Until recently, comparatively little effort had been expended in developing useful methods for determining mercury in most environmental samples (a voluminous literature exists for determining mercury in solid and rocks). The difficulties of mercury analyses are formidable because of the range of samples (air, blood, fish, hair, rocks, soil, sediment, fresh and sea water, and urine), because of the limit of detection demanded (less than about 1 ppb), and because of the need to detect total mercury, as well as amounts of different species of mercury. Finally, in most instances, a further demand was imposed--convenience for routine use.

Essentially, six types of analytical methods are or will be used to determine mercury in environmental samples: colorimetric methods, flameless atomic absorption spectroscopy (FAA), neutron activation analyses, selected emission methods, gas chromatography, and such other techniques as mass spectroscopy, X-ray spectrophotometry, X-ray absorption, polarography, radiometry, and electron spectroscopy.

Colorimetric Methods using dithizone-mercury compounds were developed as early as 1925 that would detect about 0.5 μg. The procedure consists in acidifying a finely powdered sample with bromine-sulfuric and hydrobromic acid mixture, adjusting the pH to 4 (or perhaps less), and adding a solution of dithizone in *n*-hexene. The dithizone-mercury complex is extracted into the hexane, separated, and the concentration of mercury determined colorimetrically in comparison with standards. Mainly, inorganic mercury concentration is determined by this method, though the method has the advantage of being applicable to field work.[8]

Flameless Atomic Absorption Spectroscopy takes advantage of the facts that elemental mercury is volatile and that various mercury compounds can be converted into elemental mercury. The vapor is swept into a tube, ground-state atoms absorb light from an ultraviolet lamp, and the decrease in intensity is related to concentration. Any compound, such as toluene, benzene, or SO_2, that has a significant ultraviolet absorption will interfere. This type of interference is avoided by amalgamating the mercury in aqueous solution with a silver screen. The dry screen is heated in an rf heating coil and the mercury vapor that is released is detected by means of a mercury-vapor detector.[9] Various procedures differ in how the mercury sample is digested, and there are obviously differences in detection limits. For biological samples (blood, urine, etc.), the detection limits appear to be 0.n-2 x 10^{-9} g/g sample[1] or 0.0n-0.22 ppb. A novel modification--the use of a membrane probe--permits analysis of free and total mercury, and dimethyl mercury has been detected.[10] With this easily constructed probe, the lower limit of detection was 4 x 10^{-10} g.

Neutron Activation Analysis is being used to determine mercury concentrations in sediments and water to a level of 0.05 ppb. A 20-mℓ sample is irradiated in a sealed quartz tube (1 MW , 4 hr) and two isotopes (^{197m}Hg, 24-hour half life; ^{197}Hg, 65-hour half-life) are produced. The isotopes isolated with carrier mercury salts are then reduced to elemental form, and the activity of ^{197}Hg is counted. Specific identification is confirmed by carrier isolation and the characteristic photon spectrum.[2] Neutron activation analysis is more expensive and slower than FAA. Unless modified, they lack specificity (inorganic versus methyl mercury, etc.).

Emission methods and conventional atomic absorption have had comparatively high detection limits, 2.5 and 10 ppm, respectively.[10] Modifications in emissions have and will be made, however, that should enhance limits of detection. For example, an emission spectrophotometry method that used an rf helium plasma has a working range of 1-1000 ppb (10-mℓ sample).[11]

Gas chromatography is a method that many predict will be used to give useful diagnostic information concerning the concentrations of significant species.

32.5 Mercury Determination by FAA

Several procedures are available for analysis of mercury by FAA spectrophotometry, including those described by Hatch and Ott,[12] The Dow Co.,[13] and Braman.[10] Apparatus for the first two is commercially available (Note 5); the apparatus for the last is easily fabricated.

The general procedure consists of three steps: digestion, volatilization, and detection. Solid substances (sludge and tissue) must be digested and mercury must be converted to a form that is easily reduced. Samples must also be free from organic contaminations that absorb at 2537 Å, and digestion is necessary to remove organics. Insoluble mercury compounds are solubilized with acidified permanganate.

The digested sample is reduced (typically with stannous chloride or hydrazine, though Braman[10] used sodium borohydride) and the elemental mercury thus formed is volatilized into a 10-cm analysis cell.

The sample absorption is determined at 2537 Å and the concentration is determined by comparison with appropriate standards.

Procedure

(a) Apparatus

An FAA apparatus is illustrated in Figure 32-2, kits are commercially available (Note 5).

(b) Reagents

Distilled Water. See Section 8.4.

Hydroxylamine Solution. Dissolve 10 g of reagent-grade hydroxylamine hydrochloride in 90 mℓ of distilled water and place in a dropper bottle.

Reducing Solution. Dissolve 10 g of reagent-grade stannous chloride ($SnC\ell_2 \cdot 2H_2O$) in 20 mℓ of warm conc. hydrochloric acid. Add 80 mℓ of distilled water. The solution is good for a week (Note 6).

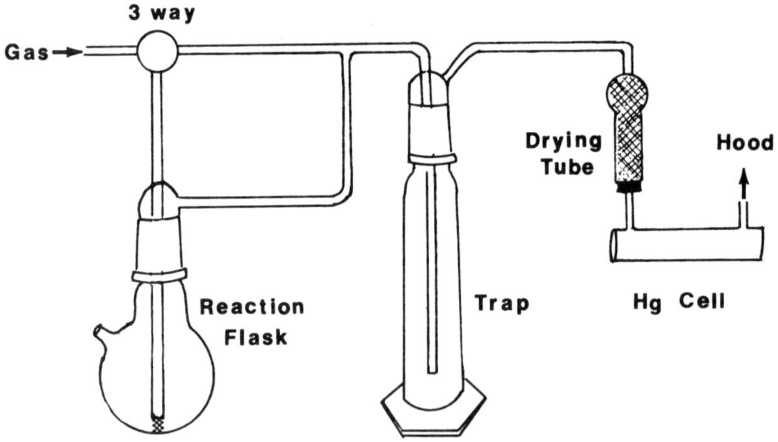

FIG. 32-2

Schematic representation of apparatus for determination of mercury by flameless atomic absorption spectroscopy.

Dilute Sulfuric Acid. Carefully pour 25 mℓ of conc. sulfuric acid (96%) into distilled water and dilute to 100 mℓ.

Mercury Standards. Stock solution (0.1 N). Dissolve 8.55 g of reagent-grade $Hg(NO_3)_2 \cdot H_2O$ in a mixture of 2 mℓ conc. nitric acid (69%) and 100 mℓ of distilled water. Dilute to 500 mℓ. Standard solution (200 ppm). Acidify 50 mℓ of 0.1 N stock solution with 5 mℓ of dilute sulfuric acid and dilute to 250 mℓ with distilled water. Standard solution (10 ppm). Acidify 5.0 mℓ of the 200-ppm standard solution with 5 mℓ of dilute sulfuric acid and dilute to 100 mℓ with distilled water. Working standard (10 ppb). Acidify 1.0 mℓ of the 10-ppm standard solution with 5 mℓ of dilute sulfuric acid and dilute to one liter. Prepare fresh daily.

Permanganate Solution (6%). Dissolve 30 g of reagent-grade potassium permanganate in 400 mℓ of distilled water with warming and stirring. Then, dilute the solution to 500 mℓ.

Sulfuric Acid, Conc. Use 96% reagent-grade acid.

Aqua Regia. Do not store this acid. Cautiously add 10 mℓ of conc. reagent-grade nitric acid (69%) to 30 mℓ of conc. hydrochloric acid (37%). Stir and allow to stand for 5-10 minutes before use.

(c) Sample Preparation[13,14]

Natural Waters. Add 1 mℓ of dilute sulfuric acid and 1 mℓ of permanganate to a 50-mℓ sample in a 100-mℓ beaker. Cover the beaker with a watch glass and boil for a few seconds.

Allow the sample to cool, then analyze. For samples that have high mercury levels (>5 ppm), use a 5-mℓ sample and dilute to 50 mℓ.

Sediments. Exclude sticks, stones, leaves, and other large particulate material from the sample. Accurately weigh a 5-g sample into a 100-mℓ beaker. Cautiously add 20 mℓ of aqua regia, and stir (hood). Boil for one minute, then allow the sample to cool. Quantitatively transfer the liquid to a 100-mℓ volumetric flask and dilute to the mark with distilled water. Use a 50-mℓ sample or suitable aliquot in the analysis. It may be necessary to use a dry weight basis (cf. Section 33.3).

Biological Tissue. Various sampling procedures are available; the procedure with fish is offered as an example. Weigh the specimen ("as is" or "wet weight"). Remove the complete fillet and carefully remove the skin from the fillet. Use a household blendor to convert the fish into a homogeneous paste. Transfer the homogenized sample to a tared, labeled bottle with pertinent information (species, weight, location and date of catch, etc.).

Digest a 0.5-g homogenized sample as follows. Accurately weigh the sample in a tared, 100-mℓ volumetric flask. Make certain that the sample is in bottom, not the neck of the flask. Carefully add 5 mℓ of conc. H_2SO_4. Cover the flask with a beaker and warm on a steam or in a boiling water bath for about 15-30 minutes to obtain a solution or homogeneous dispersion. Cool the solution in an ice-water bath for 15 minutes, and while it is in the bath, cautiously add 15 mℓ of 6% $KMnO_4$. Cautiously swirl the flask, and mix the contents. Allow the flask to stand at room temperature until rapid frothing stops (about 30 minutes).

Continue the digestion. Heat the flask on a steam bath until all frothing stops and foam disappears. Stop heating the flask, and cautiously add 5 mℓ of permanganate solution. Heat the contents just to the boiling point, than analyze the sample.

(d) Analysis

Prepare the FAA assembly. Place fresh magnesium perchlorate in the drying tube, and adjust the flow of compressed air or nitrogen through the gas cell (Note 7). Turn the three-way stop-cock to the bypass position with an empty sample bottle in position, and stabilize the baseline.

Add the digested sample to the assembly and treat. Dispel color due to excess permanganate with a few drops of hydroxylamine solution. Wash the sample into the aeration vessel, and dilute to standard volume (60 mℓ) with distilled water. Add 2 mℓ of reducing solution and attach the sample bottle. Change the three-way stop-cock to the aeration position. The mercury peak should appear very quickly. Note the maximum absorbance on the recorder.

Prepare the system for the next sample by flushing with air or nitrogen until the baseline is stabilized (perhaps two minutes at three liters/minute). Add a new sample flask, and analyze the next sample.

(e) Calculations

Standarize daily using 1-,3-,5-, and 10-mℓ additions of 10-ppb standard. Use the pertinent equation, together with the number of micrograms of mercury as determined from the daily calibration plots:

$$\text{water:} \quad \text{ppm Hg} = \frac{\mu\text{g Hg}}{\text{m}\ell \text{ sample x sp. gr.}} \text{ x } \frac{\text{dilution}}{\text{aliquot}} \tag{32-4}$$

$$\text{sediment:} \quad \text{ppm Hg} = \frac{\mu\text{g Hg}}{\text{g sample}} \text{ x } \frac{\text{dilution}}{\text{aliquot}} \tag{32-5}$$

$$\text{fish:} \quad \text{ppm Hg} = \frac{\mu\text{g Hg}}{\text{g sample}} \tag{32-6}$$

NOTES

1. For comparison, per capita fish consumption varies with the country from 84 g/day (Japan, 1961), to 56 (Sweden, 1967), to 17 (U.S.A., 1967) to 16 (Italy, 1966-1967).[1]
2. One factory at Minamata Bay had used mercury as a catalyst in acetic acid synthesis as early as 1932. The production of polyvinylchloride (PVC) using mercury-containing catalysts increased substantially after 1948 and first cases of poisoning were noted in 1953. Persons who were poisoned postnatally in Manamata had high fish and/or shellfish consumption, and though the mercury levels at the time are uncertain, the mean level was about 40 mg/kg wet weight, and about twice this level during the outbreak. The levels of methyl mercury are unknown.[1]

3. Other choices might be S-adenosylmethionine (SAM) and N^5-methyltetrahydrofolate derivatives. The first involve transfer of a methyl group as a carbonium ion CH_3^+. Methyl corrinoids, in theory, can also transfer a methyl as a carbanion (CH_3^-) or as a radical ($CH_3^\bullet$).

4. cf. Ref. 1, p. 63.

5. For example, the chemicals and apparatus exclusive of detector system are sold by Utopia Instrument Co. (Joliet, Ill. 60434), by Ocli Instruments (South Norwalk, Conn. 06854), and Laboratory Data Control (Riviera Beach, Fla. 33404).

6. Some examples of stannous chloride contain too much mercuric chloride for reasonable blanks. Either obtain a better sample or extract the solution 3-4 times with 15 mℓ of chloroform. This will remove the mercury as $H_2HgC\ell_4$, but the chloroform must be removed by bubbling nitrogen through the solution.[15]

REFERENCES

1. Methyl Mercury in Fish, Nordisk Hygienisk Tidskorift, Supplementum 4, National Inst. Publ. Health, Stockholm, 1971.

2. Mercury in the Environment, Geol. Survey, Prof. Paper 713, U.S. Govt. Printg. Office, Washington, D. C., 1970.

3. L. Dunlap, Chem. Eng. News, July 5, 1971, 22.

4. C. E. Knapp, Environ. Sci. Technol., 4, 890 (1970).

5. L. Kurland, World Neurology, 1, 370 (1960).

6. J. M. Wood, F. S. Kennedy, and C. G. Rosen, Nature, 220 (1968).

7. S. Jensen and A. Jernelov, Nature, 223, 753 (1969).

8. F. N. Ward and E. H. Bailey, Am. Inst. Mining, Metall. and Petroleum Engineers, Trans., 217, 343 (1960).

9. M. E. Hinkle and R. E. Learned, U. S. Geol. Survey Prof. Paper 650-D, p. D-251, U.S. Govt. Printg. Office, Washington, D. C., 1969.

10. R. S. Braman, Anal. Chem., 43, 1462 (1971).

11. R. W. April and D. N. Hume, Science, 170, 849 (1970).

12. W. R. Hatch and W. L. Ott, Anal. Chem., 40, 2085 (1968).

13. Determination of Mercury (Method CHS-AM-70.13), June 22, 1970. The Dow Chemical Co., Midland, Michigan.

14. J. F. Uthe, F. A. J. Armstrong and M. P. Stainton, J. Fish. Res. Bd. Canada, 27, 805 (1970).

15. R. S. Braman, Personal Communication, 1971.

Section V

ANALYSIS OF SEDIMENTS AND SELECTED ORGANIC MATTER

33

CHEMICAL PROPERTIES OF SEDIMENTS

33.1 Introduction

The chemical properties of sediments that are commonly determined include: chlorinity of interstitial water, water and carbonate content, and silicon, calcium, magnesium, and trace-element content. These determinations are considered in the sections that follow.

33.2 Chloride Content of Interstitial Water

The choice of technique used for extracting interstitial water from sediments is governed by the nature of the sediment. Siever[1] has described techniques for removing interstitial water from fine-grained sediments (such as silts and clays). Interstitial water is removed easily from compacted sediments with high moisture contents simply by squeezing the sediment in a filter press. Sands and coarser-grained sediments require special techniques which fall into four categories: centrifugation,[2] desiccation,[3] gas-extraction,[2] and immiscible-liquid extraction.[2] Of these methods, the first and last are comparatively simple and apparently yield reliable results if samples are obtained and preserved properly. A centrifugation method is given here.[4]

Procedure

Place the sample (Note 1) in two-four 10-mℓ plastic test

tubes which have 2-3 pinholes in the bottom. Fit each plastic tube in a 15-mℓ Pyrex graduated centrifuge tube in such a way that there is space in the bottom for about 1 mℓ of centrifuged water. Place the fitted tubes in a clinical centrifuge and run until a sample is obtained (3-15 minutes).

Determine the chloride content of the combined samples using the potentiometric titration procedure (Chapter 5).

Chloride concentrations as low as 200-300 ppm and as high as 16,500 ppm can be determined, but the method is not suitable for sediments that contain essentially fresh interstitial water (8-20 ppm chloride).

33.3 Water Content of Sediment Samples

The water content of a sediment is often determined as a means of indicating the degree of compaction, because the water content is related to the ease of water passage through the sediment. The ease of passage is of interest because of its relation to the nature of the sediment and the processes occurring in the sediment, diagenesis, lithification, and migration of oil. The method for determining water content depends upon the nature of the sample.

With fresh, nondried samples, the water content may be determined readily by driving off the water and knowing the weight of the sample before and after the loss of water.

With dried, recent samples, the water content may be determined less assuredly through the indirect method described by Scholl.[2] This involves analyzing the weight of halogens in the dried sediment. The halogen content in the dried sediment should be related to the chlorinity in the water overlying the sediment originally, and when this is known, the water content of the original sediment can be inferred (Note 1).

Procedure

(a) Nondried Sediments

As soon as possible after collection, place samples of the sediment from the 9-10 cm level of each cure in preweighed (nearest milligram) sealed jars. The jars should be filled to capacity and free air space should be minimized. Once the samples have been brought to the laboratory, remove the jar lid and heat the sample at 105° overnight. Remove the sample from the oven, place the jar lid loosely in place, and place the sample in a desiccator to cool. Quickly weigh the cool sample and record the weight. Sample data are recorded in Table 33-1. Given these data, the water content of the samples is calculated as follows.

Water content (wet basis):

$$W_c = \frac{\text{Wt. sample (wet)} - \text{Wt. sample (dry)}}{\text{Wt. sample (wet)}} \times 100 \qquad (33\text{-}1)$$

$$= \frac{31.400 - 23.400}{31.400} \times 100 = 25.5\%$$

Water content (dry basis):

$$W = \frac{\text{Wt. sample (wet)} - \text{Wt. sample (dry)}}{\text{Wt. sample (dry)}} \times 100 \qquad (33\text{-}2)$$

$$= \frac{31.400 - 23.400}{23.400} \times 100 = 34.2\%$$

TABLE 33-1

Sample Data for Determination of

Water Content

Before Heating	
Weight of jar and sample (wet)	71.700 g
Weight of jar (empty)	40.300 g
Weight of sample (wet)	31.400 g
After Heating	
Weight of jar and sample (dry)	63.700 g
Weight of jar	40.300 g
Weight of sample (dry)	23.400 g

Sometimes, water content is expressed as the volumetric weight or porosity n (where n is given as a percentage):

$$n = \frac{100G_sW}{100 - G_sW} \qquad (33\text{-}3)$$

where G_s is the grain specific gravity and W is the water content (dry basis). Water content on a dry basis (W) can be converted to water content on a wet basis (W_c) by the relationship

$$W_c = \frac{100W}{100 + W} \qquad (33\text{-}4)$$

(b) <u>Dried Recent Marine Sediments</u>[2]

Prepare the dried marine sediment. It is essential that the material be pulverized and thoroughly mixed because halide ions tend to diffuse to an outside layer of a core during the drying process. Dry the sample at 110° overnight to drive off any remaining water.

Conduct the analysis as follows. Accurately weigh (to the nearest milligram) a 1-g sample of the sediment in a 100-mℓ beaker, and extract the halide ions as follows. Pour 20 mℓ of hot (near boiling) distilled water on the sediment sample, mix thoroughly, and carefully decant the liquid into a 100-mℓ graduated cylinder. Repeat the washing with four 20-mℓ portions of hot distilled water (Note 2) and filter off the sediment after the last extraction, being careful to save the water.

Determine the halogen content by titration of the sample with silver nitrate (Note 3).

Calculations. The calculation of the water content involves four steps:

1. Calculate the weight of halogens present in the sample:

$$\text{Wt. of halogens} = \text{milliliters of } AgNO_3 \times \text{molarity of } AgNO_3 \times 0.03545 \text{ g} \quad (33\text{-}5)$$

2. Determine or know chlorinity and salinity of the water in contact with the original sediment.

3. Calculate the weight of water present in the original sample:

$$\text{Wt. of water} = \frac{\text{Wt. of halogens} \times S^{o}/oo}{C\ell^{o}/oo} \times 100 \quad (33\text{-}6)$$

where S^{o}/oo and $C\ell^{o}/oo$ represent the salinity and chlorinity of the overlying water (step 2).

4. Calculate the water content (dry basis):

Water content (dry basis)

$$= \frac{\text{Wt. of water present in original sample}}{\text{Wt. of dry sample}} \times 100 \quad (33\text{-}7)$$

Sample data are given in Table 33-2.

TABLE 33-2

Sample Data for Water Content in Dried Sediments

Weight of sample (dry)	1.050 g
Milliliters of 0.01989 M $AgNO_3$	2.74 mℓ
Weight of halogens	1.94×10^{-3} g
Original chlorinity	$11.47^{o}/oo$ (obs)
Original salinity	$20.73^{o}/oo$ (obs)
Weight of water	0.350 g
Water content (dry basis)	33.3%

33.4 Carbonate Content of Sediments

In this procedure, a sample of dried, solid marine sediment is treated with an excess of acid in a closed system of known volume. The corrected pressure increase is a measure of the carbon dioxide produced, which is directly related to the amount of carbonate present in the original sample:

$$CO_3^{2-} + 2H^+ \rightarrow H_2O + CO_2\uparrow \quad (33\text{-}8)$$

An all-glass apparatus may be used if very precise results are desired (Note 3). For routine determinations of fair precision and accuracy, a simple apparatus is sufficient. This consists of a U-tube mercury manometer, connected by means of rubber tubing to a 125- or 250-mℓ filter flask (Figure 33-1).

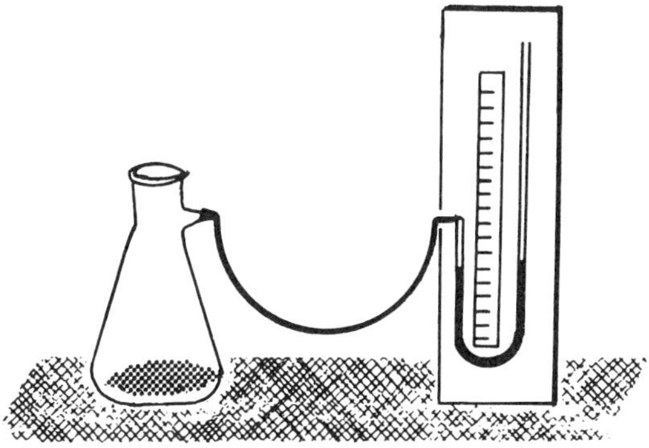

FIG. 33-1

Apparatus for determination of carbonate content.

Procedure

(a) Calibration of the System

Determine the volume of the system by using a known weight of carbonate; usually, pure sodium or calcium carbonate is used. Accurately weigh (nearest milligram) a 0.100-g sample of pure anhydrous sodium or calcium carbonate in a dry glass vial and transfer the contents to the flask. Place 5 mℓ of 0.5 M HCℓ (Note 4) solution in the flask and place the vial in the flask without the vial tipping over. Stopper the flask with the rubber stopper. Tip the flask to spill the dilute hydrochloric acid onto the carbonate sample and mix the contents by gentle swirling. Once the flask is at room temperature, record the increase in the height of the mercury column, the room temperature, and the barometric pressure. Calculate the volume of the system, as indicated in the calculation section. Repeat until successive determinations agree within 1%.

(b) Determination of Carbonate Content of a Sediment

The procedure used in calibration of the system is followed. The sample should be dried overnight at 105° and stored in a desiccator. The determination should be made with a sample of 0.2-0.5 g, depending upon the carbonate content. Duplicate determinations should be made.

(c) Calculations

Calibration of Volume of System. Two steps are involved.

First, calculate the corrected pressure of carbon dioxide, P_{CO_2}:

$$P_{CO_2} = P_{barom} + h - P_{H_2O} \tag{33-9}$$

where P_{barom} = barometric pressure, mm mercury

h = height increase of mercury, mm Hg

P_{H_2O} = vapor pressure of water at room temperature mm Hg

Second, calculate the volume of the system in milliliters:

$$V = \frac{nRT}{P_{CO_2}} \tag{33-10}$$

where n = moles of carbonate used

$$= \frac{\text{Wt. of } Na_2CO_3}{106.0} \quad \text{or} \quad \frac{\text{Wt. of } CaCO_3}{100.0}$$

R = gas constant

= 62,300 mℓ mm Hg $mole^{-1}$ deg^{-1}

T = room temperature, degrees Kelvin

= °C + 273

Calculation of Per Cent Carbonate. Calculate the corrected pressure of carbon dioxide from Eqn. (33-5), then calculate the percentage of calcium carbonate using

$$CaCO_3 = \frac{VP_{CO_2}}{T \times Wt.} \times 0.1605 \qquad (33\text{-}11)$$

where V = volume of the system, milliliters, from Eqn. (32-10)

T = room temperature, degrees Kelvin

Wt. = weight of sample, grams

0.1605 = conversion factor

Calculate the percentage carbonate as CO_3^{2-} using

$$\%CO_3^{2-} = \frac{VP_{CO_2}}{T \times Wt.} \times 0.0962 \qquad (33\text{-}12)$$

Typical data are given in Table 33-3.

35.5 Other Determinations

Methods for the determination of calcium, magnesium, silicon, manganese, copper, molybdenum, and iron in interstitial waters have been given in preceding chapters. Details for analyses of carbonate sediments are also given in Chapters 21 and 23.

No details are given for the direct determination of pH (Chapter 2). In spite of the apparent simplicity of the direct

TABLE 33-3

Sample Data for Carbonate Determination

Calibration data	
t_0	26°C
P_{barom}	763 mm Hg
h	55.7 mm Hg
P_{H_2O} at 26°C	25 mm Hg
P_{CO_2}	794 mm Hg (calc)
Weight of Na_2CO_3	0.1012 g
n	0.955×10^{-3} mole
T	299°K
V	244 mℓ (calc)
Per cent carbonate data	
Weight of sample	0.2034 g
t_0	26°C
P_{barom}	763 mm Hg
h	55.6 mm Hg
P_{H_2O}	25 mm Hg
P_{CO_2}	794 mm Hg
V	224 mℓ
T	299°K
%$CaCO_3$	46.8%
%CO_3^{2-}	28.2%

measurement, there are complex practical and theoretical problems involved. These have been described by Morris and Stumm.[5]

Determinations for soluble organic carbon, nucleic acids, and protein are given in Chapters 35 and 36.

NOTES

1. The validity of any chloride determination depends upon the suitability of the sample and the preservation of the original sample. The sample material must be compact (medium-to-fine sands and finer) and contamination by drilling must be avoided.
2. At this point, substantially all of the halide ions should be extracted, but the washings should be continued until a 1-mℓ portion of the wash water does not react with silver nitrate solution.
3. The potentiometric titration method (Chapter 5) is recommended. The wash water can be diluted with an equal volume of distilled water and two equal aliquot portions can be titrated. If this is done, the observed halogen content must be multiplied by two to get the correct value. The silver nitrate solution used in the potentiometric method should be diluted 10 mℓ to 100 in a volumetric flask.
4. Dissolve 42 mℓ of conc. HCℓ (37%) in distilled water and dilute to one liter.

REFERENCES

1. R. Siever, J. Sediment Petrol., 32, 329 (1962)

2. D. W. Scholl, Compass Mag., 35, 110 (1958).

3. D. W. Scholl, Sedimentology, 2, 156 (1963).

4. W. V. Swarzenski, Bull. Am. Assoc. Petrol. Geologists, 43, 1995 (1959).

5. J. C. Morris and W. Stumm, Adv. Chem. Ser., 67, 270 (1967).

34

HUMIC ACIDS

34.1 Introduction

A common and important constituent of most surface waters, particularly near shore, are the so-called yellow organic acids (humic substances). These materials may reduce productivity because of absorbance of light energy and interference with photosynthesis, though they appear to be beneficial at certain, lower concentrations, because of association with metal nutrients.[1-3] Their capacity for binding iron, for example is well known,[4] and may be responsible for specific influence on the growth of algae.[5]

Presumably, much of the yellow organic acids in coastal waters is primarily terrestrial in origin. Plankton decomposition products, which also impart a yellow-brown color to sea water, were termed "Gelbstoff" by Kalle,[6] but they have characteristics that are similar to the products of terrestrial origin. Both types of products comprise a heterogeneous group of polyhydroxymethoxy-carboxylic acids and quinones. Humic substances are formed from the biochemical reactions of plant and animal tissue in soil or sediment.

Humic substances can be classified into four fractions,[7] each of which contains a heterogeneous and dynamic group of substances, humins (insoluble), and three soluble fractions--

humic, fulvic, and hymatomelonic acids. The general distinction is indicated in a flow diagram summarizing their separation (Figure 34-1).

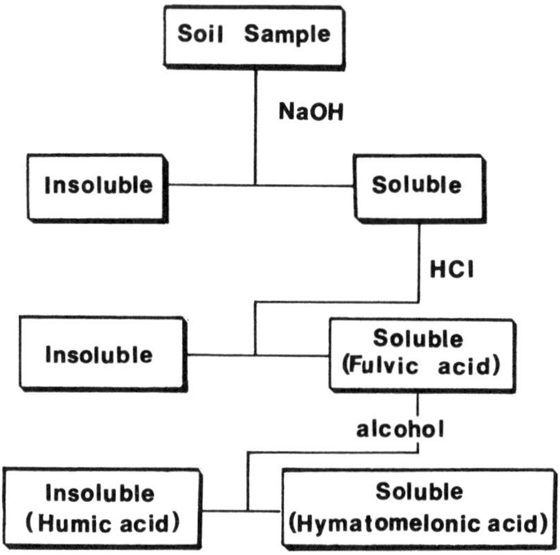

FIG. 34-1

Classification scheme for soil-organic matter.[4,5,10]

The distinction between the fractions is subject to differences of opinion among various workers. Some general remarks may be useful as a guide to the analysis of humic acids, but detailed reviews should be consulted for an adequate summary of current views. Only the soluble fractions are of interest as biologically utilizable by plankton. Some difference of opinion about the solubility of humic acid in sea water exists: according to one view, it exists as a colloidal dispersion, though low-molecular-weight fractions may be soluble in water.[8] Fulvic acid is more soluble in water than humic acid, shows fluorescence in ultraviolet light, and is less complex in structure than humic acid. Finally, some workers regard the three soluble fractions as being related compounds.

34.2 Humic Acid Analysis[9]

Humic acids may be defined operationally as three decomposition products of plant and animal tissue that comprise the base-soluble and acid- and alcohol-insoluble fractions of soil and sediments. The extraction used here follows the operational definition and is applicable to routine analysis of humic acids in rivers, estuaries, and coastal waters.

The method consists in treating a 250-mℓ sample of humic acid solution or natural water sample with glacial acetic acid then isoamyl alcohol, collecting the precipitate thus formed (after at least five hours), washing the precipitate with 95% ethanol, and then dissolving the residue in 0.5 M NaOH solution. Standard calibration curves are prepared from a standard solution of humic acid that is isolated from peat.

It has been determined[10] that humic acid samples isolated from several Florida river soils have similar properties, that the samples are representative of humic acids in west Florida rivers, and that the procedure measures humic acid contents, not tannic acid. It appears that humic acid concentrations can be determined with a 10% coefficient of variation, under realistic conditions.

Procedure

(a) Reagents

Glacial Acetic Acid. Use reagent-grade.

Isoamyl Alcohol. Use reagent-grade.

(b) Extraction of Humic Acid [9-12]

Hydrolyze one liter of peat soil sample for 48 hours with one liter of 0.5 M NaOH (Note 1) and 100 mℓ of 0.1 N $Na_4P_2O_7$. Strain the hydrolysate through two layers of cheese cloth, then through four layers of cheese cloth and centrifuge the filtrate to remove suspended base-soluble material. Discard the residue.

Precipitate the humic and hymatomelonic acids by acidifying the supernatant with an equal volume of glacial acetic acid and allow the mixture to stand overnight. The acid-insoluble material can be collected by centrifugation or less conveniently, by vacuum filtration using a Whatman No. 1 paper on a sintered-glass filter. (The supernatant or filtrate contains fulvic acid, which can be concentrated using a rotary evaporator at 85°.)

Wash the precipitate three times using 20 mℓ of ethanol to remove alcohol-soluble hymatomelonic acid fractions, which can be isolated by evaporating the ethanol extracts.

The residue is humic acid, which can be purified by resolution in 0.5 M NaOH (0.3% w/v solution) and repeating the precipitation procedure. Transfer the purified precipitate to a desiccator while moist, and allow it to dry to constant weight.

Standard Humic Acid Solution. Dissolve 40.0 mg of dry humic acid in 10 mℓ of 0.5 M NaOH solution, then dilute with distilled water to one liter (concentration, 40 ppm; pH, ~11). Dilute aliquot portions of the stock solution to give concentrations of 0.25, 0.5, 1.0, 2.5, and 5.0 ppm.

(c) Storage of Water Samples

Any samples not analyzed shortly after collection should be kept frozen to prevent decomposition or other changes.

(d) Analysis

Three steps are involved in the analysis: precipitation of humic acid, treatment, and photometric analysis.

First, treat a 100-mℓ water sample in a 250-mℓ separating funnel with 5 mℓ of glacial acetic acid, shake vigorously, then add 15 mℓ of isoamyl alcohol. After vigorous shaking, allow the mixture to stand for a minimum of five hours for complete separation and precipitation of humic acid at the interface.

Second, filter the contents of the funnel through a sintered-glass filter (medium porosity), wash the precipitate with distilled water and 95% ethanol (Note 2), and air dry. Place a 5.0-mℓ calibrated test tube in the filter flask under the tip of the filter funnel. Carefully dissolve the humic acid in 0.5 M NaOH and bring the total volume to the 5.0-mℓ mark.

Finally, determine the absorbance of the alkaline solution at 520 mμ and compare with standard solution samples analyzed in the same way.

(e) Calculation

The absorbance-humic acid relationship is linear over a concentration range of 0.5-40 ppm (1-cm cell). Most rivers on the west coast of Florida have humic acid concentrations of 1-10 ppm,[10] and the concentration in estuaries and neritic waters typically is 1 ppm or less (Note 3). Calculate the humic acid concentration from the relationship

$$\text{Humic acid (ppm)} = \text{absorbance} \times 1/m \qquad (34\text{-}1)$$

Here, m is the slope of the linear calibration plot.

NOTES

1. Dissolve 20 g of NaOH in distilled water and dilute to one liter.
2. Use polyethylene bottles and about 5 mℓ of each solvent.
3. Longer-path cells should be used for seawater samples. Also, concentration of Gelbstoff (humic material) can be effected by means of nylon columns, according to Sieburth and Jensen,[13] who obtained concentration factors of 10,000 with a recovery of 70%.

REFERENCES

1. M. Ghassemi and R. F. Christman, Limnol. Oceanog., 13, 583 (1968).

2. A. Prakash and M. A. Rashid, Limnol. Oceanog., 13, 598 (1968).

3. J. Shapiro, J. Am. Water Works Assoc., 56, 1062 (1964).

4. G. T. Felbeck, Jr., "Chemical and biological characterization of humic matter", in Soil Biochemistry (A. D. McLaren and J. J. Skujins, eds.), Vol. III, Dekker, New York, 1970.

5. G. T. Felbeck, Jr., Adv. Agron., 17, 321 (1965).

6. K. Kalle, Oceanogr. Mar. Biol. Ann. Rev., 4, 91 (1966).

7. M. M. Kononova, Soil Organic Matter, 2nd ed., Pergamon Press, Oxford, 1966.

8. W. Flagg, Acta Chem. Fenn. A., 33, 229 (1960).

9. D. F. Martin and R. H. Pierce, Jr., Environ. Letters, 1, 49 (1971).

10. D. F. Martin, M. T. Doig, III, and R. H. Pierce, Jr., Fla. Dept. Nat. Resources, Prof. Papers Ser., 12, April 1966, 52 pp.

11. R. A. Overstreet, Fla. Bd. Conserv. Leaflet, Ser. Vol. VI, Part 3, No. 4, July 1966.

12. A. P. Black and R. F. Christman, J. Am. Water Works Assoc., 55, 753 (1963).

13. J. McN. Sieburth and A. Jensen, J. Exp. Mar. Biol. Ecol., 2, 174 (1968).

35

"TOTAL" ORGANIC CARBON CONTENT

35.1 Introduction

Frequently, it is desirable to determine the organic carbon content of samples, using either particulate or dissolved material. Probably the best procedure is an instrumental one using a two-channel infrared carbon analyzer that measures inorganic carbon and total oxidizable carbon; the difference between the two corresponds to total oxidizable organic carbon. Two procedures are given here that are useful if suitable instrumentation is not available.

35.2 Oxidizable Carbon Content

The following method is designed to give a rapid estimation of the amount of oxidizable carbon in a sample. The method involves wet oxidation of carbon by acid dichromate and is based upon procedures described by Marki and Witkop[1] and Moore and Scheuer.[2]

The procedure has the advantage of rapidity, but it does have certain limitations that should be considered. It can be used as a relative method to compare the relative amounts of carbon in different samples. This is useful in assays of sediment extracts. As an absolute method, it probably gives higher results than classical procedures that involve oxidation to carbon dioxide. If a typical reference standard, glucose, is used, the results obviously are expressed in terms of carbohydrate carbon.

The last point deserves special consideration. It is equally evident that not all of the organic material present is carbohydrates. Strickland and Parsons[3] suggested that "oxidizable carbon" is a realistic measure of energy stored in a crop and that the true carbon content is within 10-20% of the oxidizable carbon value. Recent studies[4] support this suggestion to the extent that densities of phytoplankton exhibited a positive correlation with glucose concentration in many areas of a trans-Atlantic section, though in certain coastal water, the correlation was poor.

Procedure

(a) Reagents

Acid Dichromate Reagent. Dissolve 1.00 g of sodium chromate dihydrate in 20 mℓ of water, and carefully dilute to one liter with concentrated sulfuric acid (96%) by adding a little acid at a time.

Glucose Solution. Dissolve 2.50 g of glucose and a few crystals of mercuric chloride in distilled water and dilute to 100 mℓ. Store in a refrigerator and discard the solution if it becomes turbid.

Working Standard. Dilute 10 mℓ to one liter and discard after a day. Each milliliter of that solution contains 100 μg of carbon.

(b) Glassware

The glassware used must be free of dirt and grease and should be cleaned with acid-dichromate cleaning mixture.

(c) Sample Preparation

The sample must be free of chloride ion which is oxidized by dichromate. A water sample (0.5-10 liters) can be subjected to freeze-drying, but the residue must be treated with phosphoric acid (70%) and heated for 30 minutes at 100° in a sand bath. If a suitable volume is filtered to collect particulate water, use Whatman GF/C glass filter papers instead of a membrane filter. The glass filters should be heated at 500°C for 30 minutes to decompose organic matter. If culture extracts are used, the solvent must first be evaporated or subjected to freeze-drying. Finally, base-soluble organic substances can be treated directly (Section 35.3).

(d) Analysis

Mix a 0.100-mℓ aliquot with 3.9 mℓ (or suitable volume, Note 1) of dichromate reagent from a Repipet dispensor in a test tube and heat for 20 minutes at 100° in a sand bath. Dilute to 4.00 mℓ with distilled water as necessary and record the absorbance at 350 mμ.

(e) Calculations

In some instances, relative values of oxidizable carbon are desired and results are reported as per cent of the blank.

If quantitative results are desired, calculate the value using

$$\mu g\ C/m\ell = A \times F \times D.F. \qquad (35\text{-}1)$$

Here, A is the observed absorbance corrected for blank absorbance, F is the reciprocal of the slope of the calibration plot (absorbance versus μg C/mℓ, and D.F. is the dilution factor.

35.3 Base-Soluble Organic Material

The conditions that are used to extract humic acid from sediments (Section 34.2) are probably overly drastic and ambient conditions are typically milder. This procedure uses dilute aqueous ammonia to extract base-soluble organic substances (Note 2).

Procedure

(a) Reagents

Aqueous Ammonia. Dilute 7 mℓ of concentrated reagent-grade ammonia to one liter with distilled water.

(b) Analysis

Weigh about 0.5 g sample of dried (in desiccator) sediment to the nearest milligram in a 250-mℓ polypropylene centrifuge bottle (Note 3). Extract the sample by adding 150 mℓ of aqueous ammonia and constantly shaking for one hour.

Separate the extract. First, centrifuge the sample with the cap loosely screwed on. Centrifuge at 14,600 g for 70 minutes. Then, carefully decant the supernatant into a membrane filtration unit (0.45 μ filter).

Repeat the extraction procedure twice using 100 mℓ of aqueous ammonia solution;reduce the shaking period to 30 minutes. Centrifuge and filter as before.

Freeze-dry the combine filtrate and weigh the residue, which may be hygroscopic.

Determine the weight percentage of organic matter and/or determine the amount of oxidizable organic matter (Section 35.2).

NOTES

1. For example, 3.9 mℓ of reagent will oxidize up to 135 μg of carbon and a 2-cm cell may be used. Also, 10 mℓ of oxidant will be suitable for 400 μg of carbon, and a 1-cm cell can be used.

2. The procedure was used at the Branch of Organic Fuel and Chemical Resources, U.S. Geological Survey, Denver, and details were made available through courtesy of Dr. Vernon E. Swanson.[5]

3. This weight assumes about 50% organic matter, and the sample weight should be adjusted to compensate for more or less sample size.

REFERENCES

1. F. Marki and B. Witkop, Experentia, 19, 329 (1963).

2. R. E. Moore and P. J. Scheuer, 172, 495 (1971).

3. J. D. H. Strickland and T. R. Parsons, A Practical Handbook of Seawater Analyses, Fisheries Research Board of Canada Bulletin, 167 (1968).

4. R. F. Vaccaro, S. E. Hicks, H. W. Jannasch, and F. G. Carey, Limnol. Oceanog., 13, 356 (1968).

5. V. E. Swanson, Personal communication, November 1969.

36

NUCLEIC ACIDS

36.1 Introduction

Measurements of deoxyribonucleic acid (DNA) or ribonucleic acid (RNA) should serve two useful purposes: (1) to indicate the biomass and (2) to provide another growth-rate parameter.[1] The second goal was achieved with RNA, at least using marine zooplankton, but limited studies[2] of DNA distribution in the Pacific Ocean suggest the first goal is not yet attained and a quandary was encountered: Either substantial amounts of living matter were rich in DNA or this nucleic acid was associated with detritus.

One logical solution to the quandary is the study of algae under defined conditions at various stages (from growth phase, to senescence, to detritus) and measurements of DNA contents at the various stages. This solution, of course, presumes that the experiences with cultured algae can be extrapolated with some validity and that the algae have not adapted uniquely to the culture medium.

36.2 DNA Analysis

Particulate material from a given sample is collected, extracted with acetone, trifluoroacetic acid, and ethanol or ethanol-ether. Nucleic acids are separated and converted to a colored diamine complex, the concentration of which is measured by spectroscopy. One type of procedure uses a fluorimetric method,[2,3] another type uses colorimetry.[4] A modification of the last procedure is used here.

Procedure

(a) Reagents

Diphenylamine Reagent. Dissolve purified (Note 1) diphenylamine 1.0 g) in 67 mℓ of reagent-grade acetic acid and add 1 mℓ of concentrated sulfuric acid. Store the reagent in the dark. When the reagent is to be used, add 0.05 mℓ of aqueous acetaldhyde (2% by volume) to each 10 mℓ of reagent.

DNA Standard. Dissolve 15.0 mg of salmon-sperm DNA (Sigma Chemical Co.) in 1 N aqueous ammonia (Note 2) and dilute to 100.0 mℓ. Store at 5°C. Each milliliter = 150 μg DNA. Use a micropipette and prepare standards that contain 1,2,5,10,25, and 50 μg of DNA per 50-μℓ aliquot. Dilute all samples with 1 N aqueous ammonia.

Perchloric Acid (0.5 N). Dissolve 23 mℓ of concentrated acid (70-72%) in distilled water and dilute to 500 mℓ.

Trichloroacetic Acid (10%). Dissolve 50 g of acid in distilled water and dilute to 500 mℓ.

Acetone. Use absolute and 90% (10 mℓ of water and 90 mℓ of acetone) acetone.

Diatomaceous Earth Suspension. Add 0.25 g of diatomaceous earth to 100 mℓ of distilled water. Shake thoroughly just before use.

(b) Filtration and Storage of Samples

Use the procedure of Holm-Hansen and co-workers.[2] Filter water samples (Note 3) through a 0.45-μ membrane filter that has been covered with 1 mℓ of diatomaceous earth sample. Use gentle suction (one-third atmosphere). After filtration is complete, use a pair of tweezers to place the filter in the bottom of a 12-mℓ conical centrifuge tube, cover with Parafilm, and store in a freezer. Meanwhile, prepare a blank sample by filtering 25 mℓ of sea water, and treating the filter as above.

(c) Preliminary Extraction Procedure

Extract all samples and blanks and omit all DNA standards. Follow the extraction sequence listed in Table 36-1. After each addition, agitate the mixture with a vortex mixer, taking care not to lose the contents of the tube. Allow the tubes to stand for about 20 minutes, and centrifuge (four minutes at 1500 g). Use a Pasteur pipette and carefully remove the supernatant without disturbing the solid.

TABLE 36-1

Summary of Extraction Procedure

Step	Number of extractions	Volume mℓ	Reagent
1	1	5.0	Absolute acetone
2	3-4	3.0	Absolute acetone
3	1	3.0	90% Acetone
4	1	5.0	Cold (5°C) 10% perchloric acid
5	2	5.0	95% Ethanol

(d) Nucleic Acid Extraction

Extract nucleic acids by the following hot-acid extraction procedure. Add 2 mℓ of 10% perchloric acid to each centrifuge tube. Place a marble on each tube and heat the tubes at 50° for 70 minutes. Allow the extracts to cool, centrifuge, and collect the supernatant for analysis (Note 4).

(e) DNA Analysis

Colorimetric analysis is conducted in two steps, color development and analysis. One milliliter of nucleic acid extract is treated with 2 mℓ of diphenylamine solution, mixed, and incubated for 20-30 hours at 30°C. A blank is prepared using 1 mℓ of perchloric acid. Absorbances of the samples are determined at 610 and 650 mμ, using 1-cm cells. The absorbance of the DNA extract is given by the relationship

$$A_{DNA} = A_{610} - A_{650} \qquad (36\text{-}1)$$

where A_{610} and A_{650} are the absorbances at 610 and 650 mμ, respectively.

Concentrations are determined using appropriate standard solutions. The limit of detection with 1-cm cells is about 1 μg.

36.3 RNA Analysis (Note 4)

At least two convenient procedures are available for RNA analysis. One method is ultraviolet analysis. This consists in diluting the nucleic acid extract (Section 36.2d) and measuring the absorbance (1-cm cell) at 260 and 290 mμ. The RNA content can be obtained from the relationship

$$\frac{\mu g\ DNA}{g\ sample} = \frac{(A_{260}-A_{290})}{a} \times \frac{m\ell\ sample}{m\ell\ aliquot} - \frac{\mu g\ DNA}{g\ sample} \times \frac{b}{a} \qquad (36\text{-}2)$$

Here, a and b are experimentally determined conversion factors and are about 0.006 and 0.0045 for a 3-mℓ sample. Blanks must be accurately prepared because trichloroacetic acid has substantial absorption in the critical ultraviolet region.

A second method is a colorimetric procedure[5] that uses orcinol(3,5-dihydroxytoluene) and that measures either yeast RNA or arabinose. This procedure is used here.

Procedure

(a) Reagents

Orcinol Reagent. Dissolve 0.33 g of ferric chloride ($FeC\ell_3 \cdot 6H_2O$) in 500 mℓ of concentrated hydrochloric acid (37%). Just before use, add 40 mg of orcinol for each 10 mℓ of the preceding solution.

DNA Standards. Two standards can be prepared: arabinose and yeast RNA.[1]

Arabinose Standard. Prepare a solution of 100 mg of arabinose in 100 mℓ of distilled water. Prepare a working standard by diluting 1.0 mℓ in a 100-mℓ volumetric flask (0.01 mg/mℓ). This standard does not keep well, so it should be prepared fresh as needed.

Yeast RNA Standard. Heat 100 mg of Schwartz yeast RNA in a centrifuge tube with 3 mℓ of 0.5 N $HClO_4$ (Section 36.2a) at 90°C for 15 minutes. Most of the solid should dissolve, but centrifuge the suspension. Dilute 0.3 mℓ of the supernatant with 0.5 N perchloric acid in a 100-mℓ volumetric flask (0.1 mg/mℓ). Store at 0.5°C.

(b) Analysis

Mix 1 mℓ of RNA extract (Section 36.2d) with 2 mℓ of orcinol reagent in a test tube. Use 1 mℓ of 0.5 N perchloric acid as a blank. Cover the tubes with marbles and heat for 15 minutes at 100°. Cool the mixture to room temperature and measure the absorbance at 660 mμ (1-cm cell).

Use appropriate standards of yeast DNA (up to 25 μg) or arabinose standard solutions.

36.3 Protein Assay

The Folin protein assay or Lowry method[6] can be used to assay the protein content of residues from the nucleic acid extraction. The reaction of the reagents with protein to produce color occurs in two steps: (1) reaction with copper in alkali and (2) reduction of a phosphomolybdophosphotungstic acid (Folin-Ciocalteau) reagent. The copper is prevented from precipitation by means of a soluble tartrate complex. The details of the reaction are given by Lowry and co-workers.[6] More recent reports have called attention to the interference of commercial buffers that develop color in the absence of protein.[7,8]

Procedure

(a) Reagents

Folin-Ciocalteau Reagent. Purchase 2 N reagent (e.g., Fisher Chemical Co.) and dilute 100 mℓ with an equal volume of distilled water.

Copper Tartrate Reagent. Prepare the following reagents separately. Reagent A. Dissolve 20 g of Na_2CO_3 in a liter of 0.1 M NaOH. Store in a polyethylene bottle. Reagent B. Dissolve 0.5 g of $CuSO_4 \cdot 5H_2O$ and 1 g of potassium tartrate in 100 mℓ of distilled water. Reagent C. Mix 49 mℓ of Reagent A and 1 mℓ of Reagent B; discard after one day.

Protein Standard. Use crystalline bovine serum albumin (Armour and Company, Chicago) and dissolve in 1 N aqueous ammonia.

(b) Analysis

Preliminary operations include the following. Bring standards and unknown samples (usually in triplicate) to a volume of 1 mℓ with distilled water. Prepare standards containing 0 (reagent blank), 20, 40, 60, 80, and 100 μg of protein. If a residue from nucleic acid extraction is being analyzed, dissolve the sample in 1 N aqueous ammonia (dissolution) may occur slowly; treatment with 1 N NaOH and heating for ten minutes at 90°C may be necessary).

The analysis consists of the following steps. Add 5 mℓ of Reagent C to each tube, mix, and allow the mixture to stand at room temperature for ten minutes. Add 0.5 mℓ Folin-Ciocalteau reagent, mix vigorously, and allow the tube to stand at room temperature for 40 minutes. Read the absorbance at 750 mμ. Determine the protein content by comparison with standard solutions.

NOTES

1. It may be necessary to steam-distill the diphenylamine, and the acetic acid may need to be redistilled, particularly if the reagent develops a blue color upon standing.
2. Dissolve 7 mℓ of concentrated aqueous ammonia in distilled water and dilute to one liter.
3. Use about one liter for surface samples, about four liters for deep-cast samples.
4. I am indebted to Dr. Joseph G. Cory for helpful advice in the preparation of this section.

REFERENCES

1. W. H. Sutcliffe, Jr.,Limnol. Oceanog., 10 (Suppl.), R253 (1965).
2. O. Holm-Hansen, W. H. Sutcliffe, Jr., and J. Sharp, Limnol. Oceanog., 13, 507 (1968).
3. J. M. Kissane and E. Robins, J. Biol. Chem., 233, 184 (1958).
4. K. Burton, Biochem. J., 62, 316 (1956).
5. R. B. Hulbert, H. Schmitz, A. F. Brumm, and V. R. Potter, J. Biol. Chem., 209, 23 (1954).
6. O. H. Lowry, N. J. Rosebrough, A. L. Farr, and R. J. Randall, J. Biol. Chem., 193, 265 (1951).
7. J. D. Gregory and S. W. Sajdera, Science, 169, 97 (1970).
8. L. V. Turner and K. L. Manchester, Science, 170, 649 (1970).

APPENDIX A

Some Units of Marine Chemistry

Unit - Abbreviation		Equivalent	Equivalent of Similar English Unit
		Length	
Millimeter	mm	1/25 inch	1 inch = 25.4 mm
Centimeter	cm	about 0.4 inch	1 inch = 25.4 cm
Meter	m	39.37 inches	1 yard = 0.914 m
Kilometer	km	0.62 mile	1 mile = 1.61 km
Fathom	fm	6 ft	1 foot = 1/6 meter
Ångstrom unit	Å	10^{-8} cm	
Millimicron	mμ	10^{-7} cm	
Micron	μ	10^{-4} cm	
		Volume	
Cubic centimeter	cc	0.06 cu in	1 cubic inch = 16.4 cc
Milliliter	mℓ		
Liter	ℓ	1.06 liq qt	1 quart = 0.95 liter
		Velocity	
Knot	kt	1 nautical mile (6080 ft) per hour	
		Weight	
Micromicrogram	μμg	10^{-12} g	
Millimicrogram	mμg	10^{-9} g	
Microgram	μg	10^{-6} g	
Milligram	mg	10^{-3} g (0.0154 grain)	1 grain = 64.8 mg
Gram	g	0.035 oz avdp	1 ounce = 28.35 g
Kilogram	kg	2.2 lb avdp	1 pound = 453.6 g
Ton	T	2000 lb avdp	
Metric ton		2204 lb avdp	1.1 short tons

APPENDIX B

Selected Conversion Factors

Conversion	Factor	Reciprocal	log Factor	log Reciprocal
$\mu gPO_4^{3-} \rightarrow \mu gP$	0.3264	3.064	$\bar{1}.51341$	0.48629
$\mu gP_2O_5 \rightarrow \mu gP$	0.4365	2.291	$\bar{1}.63993$	0.36003
$\mu gP \rightarrow \mu g\text{-}atP$	0.03229	30.969	$\bar{2}.50900$	1.49093
$\mu gNH_4^+ \rightarrow gN$	0.7765	1.288	$\bar{1}.89014$	0.10992
$\mu gNH_3 \rightarrow \mu gN$	0.8222	1.216	$\bar{1}.91496$	0.08493
$\mu gN \rightarrow \mu g\text{-}atN$	0.07139	14.007	$\bar{2}.86864$	1.14635
$\mu gNO_2^- \rightarrow \mu gN$	0.3045	3.284	$\bar{1}.48363$	0.51640
$\mu gNO_3^- \rightarrow \mu gN$	0.2259	4.427	$\bar{1}.35400$	0.64611
$\mu gSiO_4^{4-} \rightarrow \mu gSi$	0.3050	3.279	$\bar{1}.48430$	0.51574
$\mu gSiO_4^{4-} \rightarrow \mu gSiO_2$	0.6525	1.533	$\bar{1}.81458$	0.18554
$\mu gSiO_2 \rightarrow \mu gSi$	0.4673	2.140	$\bar{1}.66960$	0.33041
$\mu gSi \rightarrow \mu g\text{-}atSi$	0.03560	28.086	$\bar{2}.55145$	1.44849
$\mu gCu \rightarrow \mu g\text{-}atCu$	0.01574	63.54	$\bar{2}.19700$	1.80305
$\mu gMn \rightarrow \mu g\text{-}atMn$	0.01820	54.938	$\bar{2}.26007$	1.73987
$\mu gFe \rightarrow \mu g\text{-}atFe$	0.01791	55.847	$\bar{2}.25310$	1.74700
$mgMg \rightarrow mg\text{-}atMg$	0.4113	24.312	$\bar{1}.61416$	1.38582
$mgCa \rightarrow mg\text{-}atCa$	0.2495	40.08	$\bar{1}.39707$	1.60293
$mgF^- \rightarrow \mu g\text{-}atF^-$	0.5264	18.998	$\bar{1}.72123$	1.27870

APPENDIX C

Vapor Pressure of Water*

Temperature °C	Vapor Pressure mm Hg
15	12.7
20	17.4
21	18.5
22	19.7
23	20.9
24	22.2
25	23.5
26	25.0
27	26.5
28	28.1
29	29.8
30	31.6

*From LABORATORY CHEMISTRY, T. Moeller and D. F. Martin, © 1965. Reproduced by permission of D. C. Heath and Company.

AUTHOR INDEX

Numbers in parentheses are reference numbers and indicate that an author's work is referred to although the name is not cited in the text. Numbers underlined show the page on which the complete reference is listed.

A

B

C

T

U

V

W

SUBJECT INDEX

A

B

C

SUBJECT INDEX

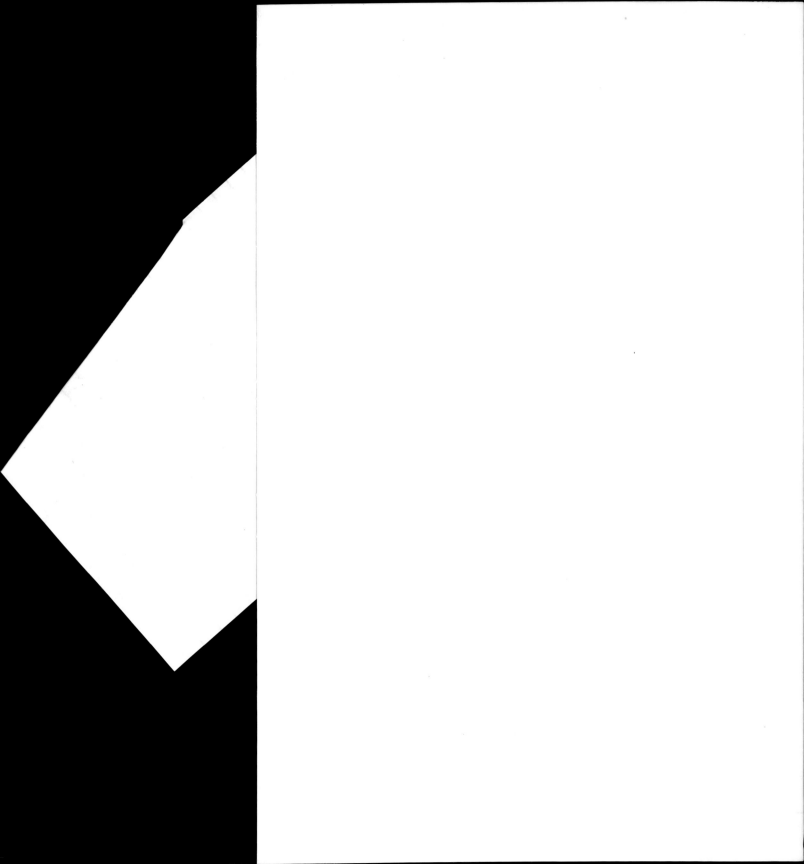

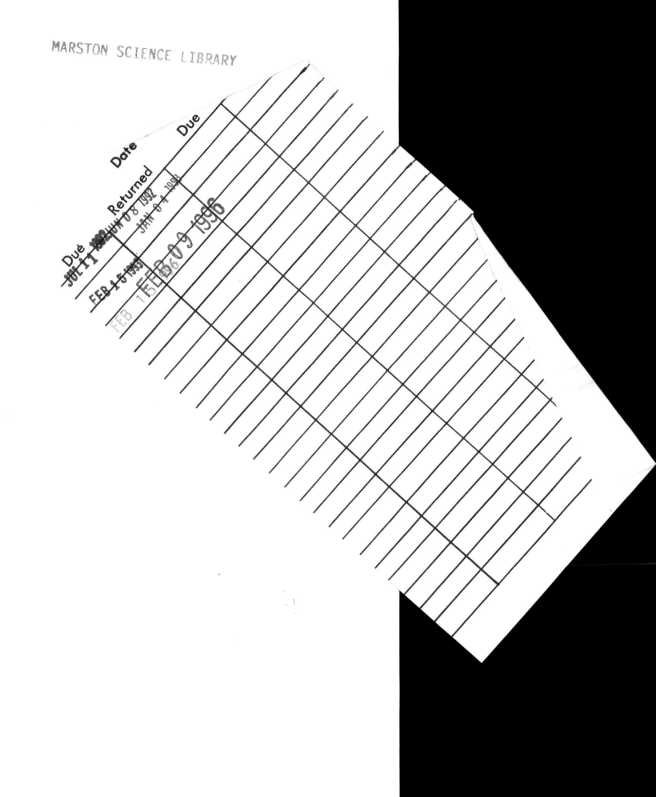